Medical Firsts

Innovations and Milestones That Changed the World

Tish Davidson

BLOOMSBURY ACADEMIC
NEW YORK · LONDON · OXFORD · NEW DELHI · SYDNEY

BLOOMSBURY ACADEMIC
Bloomsbury Publishing Inc
1385 Broadway, New York, NY 10018, USA
50 Bedford Square, London, WC1B 3DP, UK
29 Earlsfort Terrace, Dublin 2, Ireland

BLOOMSBURY, BLOOMSBURY ACADEMIC and the Diana logo are trademarks of
Bloomsbury Publishing Plc

First published in the United States of America 2023

Bloomsbury Publishing Inc does not have any control over, or responsibility for, any third-
party websites referred to or in this book. All internet addresses given in this book were
correct at the time of going to press. The author and publisher regret any inconvenience
caused if addresses have changed or sites have ceased to exist, but can accept no responsibility
for any such changes.

This book discusses treatments (including types of medication and mental health therapies),
diagnostic tests for various symptoms and mental health disorders, and organizations. The
authors have made every effort to present accurate and up-to-date information. However, the
information in this book is not intended to recommend or endorse particular treatments or
organizations, or substitute for the care or medical advice of a qualified health professional,
or used to alter any medical therapy without a medical doctor's advice. Specific situations
may require specific therapeutic approaches not included in this book. For those reasons, we
recommend that readers follow the advice of qualified health care professionals directly involved
in their care. Readers who suspect they may have specific medical problems should consult a
physician about any suggestions made in this book.

Library of Congress Cataloging-in-Publication Data

Names: Davidson, Tish, author.
Title: Medical firsts : innovations and milestones that changed the world / Tish Davidson.
Other titles: Innovations and milestones that changed the world
Description: New York, NY : Bloomsbury Academic, An Imprint of Bloomsbury
Publishing Inc., [2023] | Includes bibliographical references and index.
Identifiers: LCCN 2023013178 | ISBN 9781440877339 (hardcover) |
ISBN 9781440877346 (ePDF) | ISBN 9798216172031 (eBook)
Subjects: LCSH: Medicine—History—Miscellanea. | Medical
innovations—History—Miscellanea. | BISAC: MEDICAL / History |
SCIENCE / History
Classification: LCC R133 .D288 2023 | DDC 610/.9—dc23/eng/20230321
LC record available at https://lccn.loc.gov/2023013178

ISBN: HB: 978-14408-77339
ePDF: 978-14408-77346
eBook: 979-82161-72031

Typeset by Amnet ContentSource
Printed and bound in the United States of America

To find out more about our authors and books visit www.bloomsbury.com
and sign up for our newsletters.

Ebook ISBN needs to be updated to show the ePub ISBN rather than the ePDF.

For the next generation—Nathaniel, Abigail, Lukas, and Patrick—and for any-one anywhere who has participated in a clinical trial to advance our under-standing of the miracle that is the human body.

Contents

Chronological List
of Firsts

1. First Medical School Still in Existence, 1088
2. First Successful Cesarean Operation, 1337
3. First Hospital in North America, 1524
4. First Documented Condom Use, 1562
5. First Wheelchair, 1595
6. First American Dentist, 1630
7. First Microbiologist, 1674
8. First Clinical Trial, 1747
9. First Female Doctor with a Full Medical Degree, 1754
10. First Publicly Funded American Psychiatric Hospital, 1773
11. First Vaccine, 1796
12. First Successful Human-to-Human Blood Transfusion, 1829
13. First University-Trained African American Doctor, 1837
14. First Hypodermic Needle, 1844
15. First Use of Anesthesia in Surgery, 1846
16. First International Health Organization, 1851
17. First Epidemiological Study, 1854
18. First Use of Antiseptic Spray, 1865
19. First Successful Brain Tumor Operation, 1879
20. First Baby Incubator, 1880
21. First Steps toward Universal Health Care, 1883
22. First Cancer Immunotherapy Treatment, 1891

Thematic List of Firsts

DEVICES

DIAGNOSTICS AND RESEARCH

DRUGS AND TREATMENTS

INSTITUTIONS AND ORGANIZATIONS

PEOPLE

PROCEDURES

OTHER

Preface

Over the span of recorded history, both accidental and intentional events and discoveries have changed the practice of medicine. No book can include every medical first. The 60 medical firsts in this book were chosen because they cover diverse historical periods, countries of origin, and medical fields and because they have had an impact, not only at the time they occurred but to the present day.

Each entry begins with a discussion of the medical first itself, the person or team behind it, and the events surrounding it. Because no first occurs in a vacuum, this is followed by a historical context and impact section that outlines the social and medical context of the discovery and the impact that the discovery or innovation has had on medical care. Additional resources so that readers may learn more and better understand the value of the medical milestone are listed at the end of each entry.

The firsts in this book occurred between 1088 and 2012 with a special entry on future firsts—ideas that are under development but continue to face obstacles before they are ready for medical use. Of course, some firsts occurred before the earliest date in this book, but those events tend to be poorly documented. Many arose in China and India and were unknown to Western physicians. For example, healers in China used material from smallpox scabs to inoculate people against the disease long before Edward Jenner was credited as the originator of the first smallpox vaccine. Who really used an inoculation technique first? We probably will never know, but wherever possible, early attempts have been acknowledged in the historical context section of each entry.

The comic but classic way of representing a new breakthrough is with a light bulb switching on over the head of the discoverer. That is not how these medical firsts occurred. Only one first in this book—the discovery of X-rays—could be considered a light bulb moment of discovery. Even then, the discoverer, Wilhelm Roentgen was not sure what he had found. Fortunately, he recognized and reported the usefulness of his mystery rays. Suddenly physicians could see inside the body without surgery. Within a few weeks the approach to diagnosis had changed forever. At the other extreme, Austin Moore, who performed the first successful joint replacement, knew that his surgery would be a medical first. To make sure he got credit, he hired a professional cameraman to film the procedure.

Many of the medical firsts in this book were selected because they represent the triumph of persistence over obstacles. James McCune Smith, an excellent student, had to go to Scotland to attend medical school. He was Black, and in America in the 1830s, no medical school would admit him. Women had to overcome a similar problem. The first woman to get a medical degree from an American medical school was admitted only because her application was thought to be a prank. The school never expected her to show up. She did, and stayed to graduate.

Some researchers are to be admired for overcoming technical difficulties with patience and persistence. Frederick Sanger spent 10 years determining the amino acid sequence of insulin and another 10 years decoding the DNA genome of a tiny virus. The reward for his persistence was two Nobel Prizes. Many of the firsts in this book, however, required equal effort but resulted in either outright ridicule or delayed recognition of their impact.

Some medical firsts were chosen because they presented interesting ethical questions. Today no one considers a kidney transplant to be an ethical issue, but the doctor who performed the first one sought both medical and spiritual advice before he proceeded. The development of oral contraceptives also challenged the legal and ethical thinking of the time. In the United States, the pill remained illegal in some states until 1965. Legalization of abortion and legalization of euthanasia are examples of ethical dilemmas physicians still face. Ethical questions are put in context in the impact section of each entry.

No single picture can be drawn of the people in this book who first made discoveries and innovations that saved lives and changed the practice of medicine. Dedicated scientists, lucky researchers, the over-educated and the under-educated, scoundrels (including one murderer), and the morally upright—whatever their personal qualities, they all have one thing in common. They created devices, drugs, and procedures to reduce pain and suffering, save lives, and raise medical practice to a new level.

Acknowledgments

I am grateful for the help and support of some caring people who listened patiently to my stories about medical firsts, made constructive suggestions, read and reread the manuscript, and calmed my fears that the book would never be finished. Thank you Helen Colby and Scott Davidson. Special thanks to my talented editor Maxine Taylor. Her guidance, thoughtful suggestions, and patience helped shape and improve this manuscript. Together we birthed this book. Finally, I would also like to express my appreciation for the production team at ABC-CLIO. These behind-the-scenes people have once again worked their magic and turned pages and pages of manuscript into an attractive, accessible book.

Introduction

Today our culture seems obsessed with being the first, discovering the first, or inventing the first of almost anything—no matter how trivial—in an effort to gain instant celebrity and fortune. But when we look back at the discoverers, inventors, and innovators in this book who are remembered today for achieving real advances that changed the world of medicine and improved many lives, it is hard to find one who was motivated by the desire for fame or wealth. Some ended up enjoying both when their "first" was acknowledged. Others did not live to see the impact of their work.

Some medical firsts came about because of deaths—lots of deaths—that motivated the innovators to fight on the side of life. James Lind, a surgeon on a British naval ship, routinely saw half the sailors on long voyages die from illness. He found this unacceptable. To explore why so many sailors were dying, he created the first clinical trial.

Stéphane Tarnier, a French obstetrician, lived in a time when almost every premature baby died. In 1878, he visited the Paris zoo and saw the eggs of fragile, exotic birds hatching in a warm incubator. Could the death of babies be prevented the same way? From this thought came the first baby incubator.

Margaret Sanger found it infuriating that women had no control over their fertility. She blamed her mother's early death on her 18 pregnancies that resulted in 11 children. Sanger was not a scientist, but she crusaded for fertility control until she found three like-minded people—a funder, a researcher, and a physician—with the skills to attack the problem. The result was the first oral contraceptive and a huge improvement in women's health. All these people were motivated to solve problems that mattered to them for personal reasons, not for fame, not for fortune.

On the other hand, some medical firsts came about accidentally and could have gone unnoticed without sharp observation, curiosity, and thorough reporting. On November 8, 1895, German physicist Wilhelm Roentgen accidentally stumbled on X-rays while studying a different phenomenon. He recognized the value of his surprise discovery and announced it to the world on December 28. Eight days later the first diagnosis by X-ray was made, and medicine was changed forever.

Irving Selikoff and Edward Robitzek were testing a drug to treat tuberculosis when they noticed that patients treated with this new drug became more communicative and cheerful. Instead of dismissing this as unimportant to their study, they reported the effect in a journal article that, by chance, was read by a psychiatrist who had no interest in tuberculosis but was interested in treating depression. The result—the first drug to treat major depression.

In 1964, Jerome Horwitz, an American organic chemist, did some experiments to see if the drug azidothymidine (AZT) would kill cancer cells in mice. It did not. He wrote a journal article about this failure and put the drug aside. Twenty-three years later, during the AIDS crisis, researchers were searching for drugs with antiviral activity. AZT fit that profile. It became the first successful treatment for HIV/AIDS infection.

While accidental discoveries are exciting and memorable, most medical firsts were driven by curiosity, hard work, and persistence. Frederick Banting sold everything he owned in order to research insulin. Frederick Sanger spent 10 years decoding the DNA sequence of a tiny virus. The team that cloned Dolly the sheep made 277 attempts before they succeeded. These researchers won prizes and became famous, but that was not what drove them. They did what they did because they had an idea they wanted to test or a question they wanted to answer.

Some of the people in this book were undeniably brilliant, but brilliance is not required to change the world. John Gurdon was told by his teachers at Eton that he was not intelligent enough to study science. He was admitted to Oxford University only because of a web of family connections and favor trading. Despite his supposed lack of ability, he went on to have a brilliant career and win a Nobel Prize for his work in stem cell research.

Godfrey Hounsfield, inventor of the CT scanner, was a loner. He left school at age 16. Years later, he attended a trade school. He never went to a four-year college, but he won a Nobel Prize.

What did these innovators have in common? They asked questions, and they cared about the answers. They were creative thinkers and careful observers, persistent, and unafraid to work long and hard, to fail, and to try again. The desire for fame and fortune did not figure in their calculations, but today we still remember and admire them. Their diverse experiences showcase the many routes to discoveries and innovations that change the world.

Medical Firsts

1. First Medical School Still in Existence, 1088

The University of Bologna in Bologna, Italy, is the oldest continuously operating medical school and university still in existence as of 2023. The university traces its origins to an educational institute founded in 1088. In 1158, it received a charter from the Holy Roman emperor Frederick Barbarossa (1123–1190), also known as Frederick I. Scholars think that the word "universitas" was applied for the first time to this educational institution.

Many details of the organization of the university have been lost, but it appears that students and teachers formed what would today be called "interest groups" or "collectives" of artists, theologians, law students, medical students, and so forth. These collectives gradually transformed into various schools within the university. The University of Bologna School of Medicine and Surgery (its modern name) was formally recognized around 1200.

The medical school drew students from all parts of Italy as well as regions that are now part of Spain, the Netherlands, Hungary, and the regions bordering the Adriatic Sea. These students formed associations or guilds based on the region of Italy they called home or their nationality. The associations provided housing and secured legal protection for foreign students. By 1300, the university as a whole had about 10,000 students.

The course of study at the medical school took four years to complete. It consisted of medical lectures along with the study of grammar, rhetoric, logic, geometry, arithmetic, music, and astronomy. The physiology lectures at that time emphasized maintaining or restoring balance among the four humors: phlegm, blood, black bile, and yellow bile. Disease was thought to be sent by God, caused by an alignment of the heavens, or brought on by exposure to bad air.

The Bologna medical school emphasized knowledge of anatomy. Unlike many medical schools founded later, it used standard Greek and Latin texts but also drew on medical books in Hebrew and Arabic. Surviving writings by some professors make it clear that early in its existence, the medical school performed human cadaver dissections—some of the information in these early writings could only have come from close, direct study of internal human anatomy. Cadaver dissection was in direct contradiction to civil and religious laws of the time. Apparently, lecturers found ways to skirt the prohibition against human dissection despite the fact that a window was installed in the lecture hall so that a member of the clergy could observe, without being seen, what medical students were being taught.

Although most students were men, there are some indications that women were permitted to attend and to teach early in the medical school's history, especially in the area of childbirth. The first confirmed female professor, also considered the first female professor in the world, was Laura Bassi (1711–1778). Bassi was a child prodigy who was educated in multiple languages by a clergyman and in science by a professor at the medical school. She was appointed to the chair in Natural Philosophy (physics) in 1732. A brilliant woman interested in Newtonian mechanics, she wrote multiple theses: 1 on chemistry, 13 on physics, 11 on hydraulics, 2 on mathematics, 1 on mechanics, and 1 on technology while giving birth to 12 children.

The University of Bologna fell on hard times in the 1800s. It was reorganized in 1860 and became a national university. Study at the modern University of Bologna School of Medicine and Surgery consists of three two-year programs. The first program teaches the fundamentals of medicine. The second teaches basic tools, clinical approaches, and communication skills. The third emphasizes clinical work and an internship.

HISTORICAL CONTEXT AND IMPACT

Medicine in the Middle Ages consisted mostly of folklore, herbalism, and superstition. The average life expectancy in the 1100s was less than 35 years; 20% of children died at birth and many more died before their fifth birthday. Most people lived in rural areas where the responsibility for treating the sick fell mainly on family members. Health care generally consisted of prayer, making patients as comfortable as possible, and waiting to see if they recovered or died. In the cities, barber-surgeons, apothecaries, and a few physicians could be called on to treat the sick by those rich enough to afford their services. Their treatments were often ineffective and sometimes harmful.

The first documented medical school, the Schola Medica Salernitana, was founded in the 800s in Salerno, Italy. The school began as an informal collection of masters and students and gradually evolved into a formal institution. It was at its most influential between the eleventh and thirteenth centuries. During that time, the school was the source of medical knowledge for most of Western Europe. Its library contained texts in Arabic and Hebrew as well as Greek and Latin. Three influential works known collectively as the Tortula texts were concerned with childbirth and treatments for women. One of these texts has definitely been identified as being written by a woman, suggesting that women taught at the school. The Schola Medica Salernitana survived until November 1811, when it was abolished in a reorganization of Italian public education.

The oldest British medical school still educating doctors today is London's Medical College of St. Bartholomew's Hospital. After multiple mergers, it is now known as Queen Mary–Barts and The London School of Medicine and Dentistry. The first medical school in the United States was established in Philadelphia in

1765 by John Morgan (1735–1789) and William Shippen (1736–1808), both physicians educated in Scotland. This school evolved into today's Perelman School of Medicine at the University of Pennsylvania.

Medical education has changed greatly since the establishment of the first medical school. Initially, most healers learned their trade through an apprenticeship with another healer or, in rural areas, through informal connections. This informal system was especially active among women who practiced midwifery.

Early medical schools had few admission criteria other than the ability to pay. Education was mainly theoretical. Students studied texts written by early physician-writers such as Hippocrates (c. 460–370 BCE) and Galen (129–c. 210) and attended lectures and classroom demonstrations. They received little or no hands-on work. Some students examined their first patient only after graduation.

As medical schools became associated with hospitals, practical training was added to the curriculum. Some schools also required an apprenticeship with a graduate physician. Even with the best intentions, physicians lacked accurate knowledge. They often performed procedures such as bleeding, purging, and treating diseases with poisons such as mercury and arsenic that were harmful to the patient.

In the United States, freestanding for-profit medical schools were common from about 1820 to 1910. Courses were not graded, even at university-associated medical schools such as Harvard. There were no educational prerequisites for enrollment until 1898, when Johns Hopkins University began requiring men (no women were admitted) who attended its medical school to first earn a bachelor's degree. Harvard followed suit a few years later.

A major improvement in medical education in America occurred after the 1910 publication of *Medical Education in the United States and Canada*, written by Abraham Flexner (1866–1959) under sponsorship of the Carnegie Foundation. Flexner, an expert in educational practices but not a physician, visited and evaluated dozens of medical schools. The *Flexner Report*, as it is called today, changed American and Canadian medical education. It criticized the poor standards of many medical schools that offered only minimal facilities, a few months of lectures, and sometimes an apprenticeship under a practicing—but equally poorly trained—doctor.

Flexner advocated for a German medical school model that combined laboratory work, rigorous science lectures, and hands-on clinical training. As a result of Flexner's report, for-profit schools were closed and inadequately equipped university medical schools were forced to either close or improve their facilities and teaching methods.

Today 155 accredited schools in the United States grant Doctor of Medicine (MD) degrees and 36 grant Doctor of Osteopathic Medicine (DO) degrees. Admission to these schools is highly competitive and requires good undergraduate grades, course work in the basic sciences, and usually a four-year bachelor's degree. In addition, applicants must submit Medical College Admission Test scores, letters of recommendation, personal essays and must get through an in-person interview.

The course of study is four years. The first two years emphasize basic science education while the next two years concentrate more on clinical training and direct patient interaction. Upon completing their degree, newly minted doctors spend an additional three to seven years in internships and specialist residencies. Licensing of physicians is done at the state level with a series of examinations usually beginning at the end of the second year of medical school. Throughout their careers, physicians must earn continuing education credits to stay current in their field and retain their licenses.

FURTHER INFORMATION

Buford, Juanita F. "Medical Education." In *Encyclopedia of Education*. Farmington Hills, MI: Cengage, 2020. https://www.encyclopedia.com/history/united-states-and-canada /us-history/medical-education

Duffy, Thomas P. "The Flexner Report—100 Years Later." *Yale Journal of Biological Medicine* 84, no. 3 (September 2011): 269–279. https://www.ncbi.nlm.nih.gov/pmc/articles /PMC3178858

"Laura Bassi and the Bassi-Veratti Collection." Stanford University Library Collection. Undated. https://bv.stanford.edu/en/about/laura_bassi

Moroni, Paolo. "The History of Bologna University's Medical School over the Centuries: A Short Review." *Acta Dermatoven APA* 9, no. 2 (2000): 73–75. http://www.acta-apa .org/journals/acta-dermatovenerol-apa/papers/10.15570/archive/2000/2/Moroni.pdf

2. First Successful Cesarean Operation, 1337

Documentation of the first successful cesarean section (alternate spelling cesarean section, also called C-section) is minimal. References to surgical removal of a baby from the womb go as far back as the Roman Empire (27 BCE–476 CE) or even earlier in India and China, but there is no evidence that these operations were ever successful in saving the lives of both the mother and the child.

In 2016, the *New York Times* published an article stating that Czech medical historians had found evidence that in 1337 in Prague, Beatrice of Bourbon (1320–1383), wife of King John of Bohemia (1296–1346), and her newborn son had both survived after a cesarean operation. If accurate, this would be the first documented successful cesarean. The Czech researchers' conclusion was based on an interpretation of two letters written at the time of the birth and a third writing that later described the event. The interpretation of these writings and the researchers' conclusion were immediately challenged in an article published a few weeks later in *Forbes*. In that article, another medical historian called the interpretation "amateurish" and incorrect.

It is well documented that Beatrice gave birth on February 25, 1337, to a live son named Wenceslaus. It is also indisputable that Beatrice, aged 17 at the time, lived another 46 years and had no more children. The absence of additional pregnancies was unusual for such a young woman in a time when there was no effective birth control. At issue is whether Beatrice actually had a cesarean operation. The writings apparently indicate that she experienced a difficult labor, that Wenceslaus was taken from Beatrice's body, and that the wound healed. Nowhere is it explicitly stated that the child was *cut* from the womb. The issue of whether this was the first successful cesarean remains a topic of debate among academics.

Prior to 2016, some historians thought the earliest cesarean in which both mother and child survived occurred in the 1500s. However, this event is not universally accepted as true. Jakob Nufer of Sigershaufen, Switzerland, is said to have performed a cesarean on his wife after multiple midwives were unable to deliver their baby. Nufer was a pig farmer and may have had some concept of anatomy and birthing that could have helped him operate on his wife. Historians, however, question the accuracy of this story for two reasons. First, the story was not recorded until 82 years after it supposedly happened. Second, there is no proof that the baby survived. Nufer's wife did survive and went on to have five more children.

It may never be clear whether either or both of the women experienced a successful cesarean. In both cases, given the state of medical knowledge at the time, it would have been an extremely rare event.

HISTORICAL CONTEXT AND IMPACT

In popular culture, the word "cesarean" is thought to come from the idea that Roman statesman Julius Caesar (100 BC–44 BC) was born through this procedure. Historians dispute this interpretation. No writings from the time mention this aspect of Julius Caesar's birth, but records do show that his mother survived and lived for many more years. Survival of both mother and child in a cesarean birth was something so highly unusual in the Roman Empire that it would almost certainly have been recorded. Historians believe instead that the word "cesarean" most likely derives from the Latin *caedare*, which means "to cut."

The original purpose of a cesarean section had nothing to do with saving either mother or child but instead had to do with religious law. In ancient Rome, a law called Lex Caesarea prohibited pregnant women from being buried. The initial use of a cesarean operation was to remove a fetus from a dead or dying mother in order to comply with this law. The mother was not expected to live. Normally, this operation was performed either after the mother was dead or when it was clear that she soon would be. Waiting until the mother was *in extremis* normally caused the baby to die from lack of oxygen, but it allowed fulfillment of the religious law that mother and child were required to be buried separately.

Gradually, the idea of saving the baby's life, or sometimes the mother's life, came to be associated with cesarean section. Often, the operation was done when it was clear that the woman's birth canal was too small to accommodate the size of the baby's head or when the baby was stuck in a transverse position. Even then, the operation was to be avoided if possible because the mortality rate was as high as 85%. Women died during the operation, due to shock and blood loss, or after the operation, from infection. Early surgeons did not suture the uterus shut after operating, believing that the organ would heal itself. And these operations were performed without anesthesia until the mid-1800s, although alcohol was sometimes given to help dull the woman's pain.

Despite the high death rate, surgeons sometimes had no choice but to perform the operation. The alternatives were to either let the woman die in labor or use instruments to kill and dismember the baby in the womb and extract it piece by piece, something even the most hardened doctor would find difficult to accept. At least with a cesarean, there was a chance the baby would live.

Cesarean operations were done all over the world. In England, James Barlow (1767–1839) of Blackburn performed the first well-documented cesarean. Barlow was a qualified doctor and the author of *Essays on Surgery & Midwifery: With Practical Observations, and Select Cases*, in which he described three cases where he had performed cesarean sections, all on women who had deformed pelvic bones that prevented them from delivering normally. In two cases the woman died, and in the third, the baby died but the mother lived another 28 years.

In the United States, the first cesarean where mother and child both survived was performed in 1794 on Elizabeth Bennett by her husband Dr. Jesse Bennett (1769–1842) after the attending physician refused to perform the operation on moral grounds. In 1879, Dr. Robert Felkin (1853–1926), a British traveler, observed a cesarean performed on a young woman in Uganda. The baby survived and the mother appeared healthy 11 days after the operation when Felkin left the community.

As doctors learned more about infection control and anesthesia, cesareans became safer and were performed more frequently. Another factor contributing to their increase was that over the years more babies were born in hospitals. In 1938, 50% of American babies were born in hospitals and the rate of cesareans was under 5%. By 1955, 90% of births occurred in hospitals, and the rate of cesareans had begun to climb. Between 1965 and 1985, the rate increased by about 400%.

By 2018, about one-third of babies in the United States were born by cesarean section. The World Health Organization recommends that no more than 15% of babies should be born by cesarean. The operation is done under anesthesia, takes about 45 minutes, and requires a hospital stay of a few days. Reasons for the increased number of cesareans include better fetal monitoring that can detect fetal distress early in labor, an increase in diabetes and obesity in young women that make them potential candidates for a cesarean, an increase in the number of births in women over age 40, more multiple births, early detection of the baby in an abnormal position, and convenience of the patient or doctor. The increased rate of cesareans has become a medical concern because the operation increases the risk of complications to both mother and baby. Maternal complications include the risk of hemorrhage, infection, reaction to anesthesia, blood clot formation, and surgical injury. Potential complications for the baby include breathing difficulties and surgical injury.

Early cesarean operations had a high death rate. The scar on this woman's abdomen suggests that she has had a previous successful cesarean surgery. (Courtesy of the National Library of Medicine)

In the United States, data from the National Center for Health Statistics show that Black women have a disproportionally high rate of cesarean births. During the period 2016–2018, nationally, 35.4% of Black women gave birth by cesarean compared to 32.7% of Asian women and 31.1% of white women. The rate also varied by region. It was highest in the South. In addition, the rate was more than double for women aged 40 and older (48.1%) compared to women under age 20 (20.0%). The increase in cesareans and the health disparities they represent had by 2018 become a matter of such concern that California began a campaign to reduce cesarean births.

FURTHER INFORMATION

de Goeij, Hana. "A Breakthrough in C-Section History: Beatrice of Bourbon's Survival in 1337." *New York Times*, November 23, 2016. https://www.nytimes.com/2016/11/23/world/what-in-the-world/a-breakthrough-in-c-section-history-beatrice-of-bourbons-survival-in-1337.html

Dewell, Jane. "Cesarean Section—A Brief History." U.S. National Institutes of Health, July 26, 2013. https://www.nlm.nih.gov/exhibition/cesarean/index.html

Killgrove, Kristina. "Historians Question Medieval C-Section 'Breakthrough,' Criticize New York Times Coverage." *Forbes*, December 13, 2016. https://www.forbes.com/sites/kristinakillgrove/2016/12/13/historians-question-medieval-c-section-breakthrough-criticize-new-york-times-coverage/?sh=532259d0270a

3. First Hospital in North America, 1524

The first hospital built in North America was the Hospital de Jesús Nazareno in Mexico City, Mexico. It was established by Hernán Cortés (1485–1547), the Spanish explorer who caused the collapse of the Aztec empire. The hospital opened in 1524 for the purpose of treating both wounded Spanish soldiers and Aztec warriors. This hospital claims to have performed the first autopsy in North America. Today the Hospital de Jesús Nazareno treats patients in a modern building, but the original hospital building and the church associated with it remain nearby and can be visited by tourists. They are some of the oldest buildings in Mexico City.

The first hospital in what is now the United States was founded in St. Augustine, Florida, in 1597 when the territory was under Spanish rule. King Phillip II of Spain (1527–1598) directed the governor of the territory, Gonzolo Méndez de Canzo (c. 1554–1622), to establish a house to care for the sick, with instructions that the facility accept only those who were poor and did not have a contagious disease. The governor saw that an establishment like this could benefit his soldiers who were stationed in the city. He financed the building by taxing the soldiers, who were then guaranteed care at the hospital, ignoring the limitations set by the king.

The hospital had barely opened when a fire swept through St. Augustine, destroying the Franciscan convent but sparing the hospital. The Franciscans must have had considerable political influence because they took over the hospital for their own use, discontinuing its care of the sick. A new hospital was built and opened in 1599. Like the first hospital, it was partially financed by a tax on the soldiers stationed in St. Augustine. The staff was made up of convicts, enslaved people, and retired soldiers.

Sometimes, a physician—known at the time as a barber-surgeon—was part of the military garrison and served at the hospital. At other times, the patients made do with whatever care the untrained staff could provide. According to the *St. Augustine Record*, "additional civilian physicians came to St. Augustine by luck, when they were captured or their ship wrecked near the city." By the mid-1600s, a second hospital had been established, and the original building became a hospital exclusively serving soldiers garrisoned in the city.

Another early hospital established outside the original Thirteen Colonies, the L'Hôpital des Pauvers de la Charité, later known as Charity Hospital, was founded in New Orleans in 1736 when the city was under French control. This hospital continued to operate until it was forced to close by Hurricane Katrina in August 2005. It never reopened. In 2019, the Louisiana State University Board of Supervisors—who own the hospital—voted to allow the building to be renovated into residential units, retail stores, and restaurants.

Some sources discount the idea of these early hospitals as American "firsts" because they were built in areas outside the original Thirteen Colonies that formed the earliest version of the United States. Even within the Thirteen Colonies, there is disagreement about which hospital was the first to be established because many hospitals evolved from institutions that originally served other purposes. Bellevue Hospital in New York City claims to be the first publicly funded hospital established in the Thirteen Colonies. This hospital grew out of an almshouse whose purpose was to shelter the poor, elderly, ill, and homeless. When construction of the almshouse was completed in 1736, it contained areas for hard labor, instruction in sewing, knitting, weaving, iron work, and leather work. It also had a second-story room that contained a six-bed infirmary that was under the supervision of a medical officer. Although expansion forced a move from its original site, Bellevue Hospital continues to serve patients today as part of the New York City hospital system.

The Pennsylvania Hospital in Philadelphia also claims to be the first hospital established in the Thirteen Colonies because it was the first to receive an official charter from the governor of the state. Like Bellevue, it traces its beginnings to an almshouse built in 1730. The almshouse provided some medical care, but its connection to the Pennsylvania Hospital is somewhat obscure. The definitive establishment of the Pennsylvania Hospital occurred in 1751 when statesman and diplomat Benjamin Franklin (1706–1790) and Dr. Thomas Bond (1713–1784), a Quaker physician, secured a charter from the governor of Pennsylvania to build a hospital specifically to care for the sick and mentally ill.

At this time, there were fewer than 200 physicians with medical degrees in the American colonies; most had received their training in Europe. The Pennsylvania Hospital opened the nation's first school of medicine in 1765 and established the first medical library in America. The hospital was what would today be called a public-private partnership. The governor's charter required that the hospital raise private donations to match or exceed the amount of money provided by the government. The Pennsylvania Hospital, renamed the Hospital of the University of Pennsylvania in 1874, still exists today as does its medical school and medical library, all of which are part of the University of Pennsylvania Health System.

HISTORICAL CONTEXT AND IMPACT

As far back as Greek and Roman times, hospitals were primarily places to treat wounded soldiers. The hospitals established by Hernán Cortés in Mexico and Gonzolo Méndez de Canzo in St. Augustine were also mainly for the benefit of soldiers who were far from family. Trading companies such as the Dutch West India Company established a few early hospitals to care for sailors who became ill far from home.

Many early hospitals developed from workhouses or almshouses. These were publicly funded accommodations of last resort for the impoverished, the elderly, the disabled, and the desperate. Other hospitals developed from shelters run by religious orders dedicated to care of the poor. The goal of these institutions was to provide a minimum level of shelter and sustenance, but inevitably many of the destitute were ill, so some of these shelters gradually evolved into institutions that provided basic medical care.

Until late in the nineteenth century, the belief was that illness was best treated at home. Doctors made house calls, delivered babies, and even performed surgery in private middle-class and upper-class homes. Day-to-day care of the sick was considered a duty of the women of the house. Hospitals mainly served an urban population and were associated with the poor and the diseased. Given the level of crowding, poor sanitation, and untrained staff in early hospitals, most people were probably better off being treated at home.

In 1839, Dr. Joseph Warrington founded the Nurses Society in Philadelphia. He wrote a book that was used to train women to care for women in childbirth as well as new mothers and their infants. Most of the women who worked for the Nurses Society cared for women in their homes. Today they would be considered private duty nurses.

The American Civil War (1861–1865) emphasized the need for nurses. Over the course of the war, a total of about 20,000 women and men on both sides of the conflict served as mainly volunteer nurses. Their success encouraged the development of professional nursing training. Women's Hospital of Philadelphia developed a six-month nurse training course in 1869. This was followed a few years later by nurse training programs at Bellevue Hospital, the Connecticut State Hospital, and Massachusetts General Hospital in Boston. By 1900, there were as many as 800 two- or three-year nurse training programs in the United States, most of them affiliated with hospitals where nurses in training could gain practical experience.

With advances in medicine and specialized medical equipment, a better understanding of germs and disease transmission, and a supply of trained nurses, care of the sick shifted from homes to hospitals. No longer were family members expected to be the primary caretakers of their loved ones.

As of 2020, there were close to one million licensed doctors and 3.8 million registered nurses (RNs) in the United States. About 36 million patients were treated annually at 6,146 American hospitals. Seventy-eight percent of these hospitals were evaluated and accredited by The Joint Commission, formerly called the Joint Commission on Accreditation of Hospitals, which assures a high standard of care. The remaining 22% were evaluated by other specialized accrediting organizations. The days when hospitals were seen as dirty and crowded—the last resorts of the ill and the poor—are long gone.

FURTHER INFORMATION

Michaels, Davida. "A Brief History of American Hospitals." American Nursing History, October 1, 2019. https://www.americannursinghistory.org/hospital-history

Oshinsky, David. Bellevue: Three Centuries of Medicine and Mayhem at America's Most Storied Hospital. New York: Doubleday, 2016.

Parker, Susan. "The First Hospital in What Is Now the United States." St. Augustine Record, March 28, 2010. https://www.staugustine.com/article/20100328/news/303289998

4. First Documented Condom Use, 1562

We will never know exactly when the first condom was used. Some historians speculate based on drawings that ancient Egyptians and Greeks used linen material to cover the glans, or tip of the penis, to prevent disease, although this interpretation is disputed. A cave painting in France from around 200 CE seems to show a man having intercourse with something on his penis, but again, the interpretation of the drawing is in the eye of the beholder.

The first well-documented description of a condom comes from the Italian physician Gabriele Falloppio (1523–1562) in the book De Morbo Gallico (The French Disease), published two years after his death. Falloppio recommended that men use a linen sheath soaked in chemicals, dried, and then tied with a ribbon so that it covered the glans. Its purpose was to protect against syphilis, not to prevent pregnancy. Falloppio claimed to have tested this condom's protective value on more than 1,000 men, none of whom developed syphilis.

Besides linen sheaths, early condoms were made of cleaned animal intestines or animal bladders that were softened with lye. Bladder condoms were sometimes referred to as "skins." To date, the earliest condom found by archeologists has come from the excavation of Dudley Castle in England. This condom, made of animal membrane, probably intestine, was preserved in an ancient latrine and has been dated to about 1640. A condom of similar age was also excavated in Lund, Sweden. It was made of pig intestine, came with an owner's manual written in Latin, and was reusable. We will never know who used the first condom, but it is clear that animal membrane condoms were used over a wide geographic area, at least among the wealthy, by the mid-1600s.

HISTORICAL CONTEXT AND IMPACT

Condoms did not begin as contraceptive devices. A major outbreak of syphilis occurred among French troops in 1494. Over the next century, syphilis spread rapidly through Europe, the Middle East, and Asia. Written records from the time indicate that this was a particularly virulent strain of the disease, much more lethal than today's variety. And there was no cure. This stimulated interest in finding ways to protect those having intercourse from becoming infected. Making condoms out of animal intestines or bladders was complicated and time-consuming. These early condoms were expensive, which limited their use to the upper class. The poor simply had to take their chances, and many died of syphilis.

Condoms have been called by a number of names—male sheaths, rubbers, jimmies, love gloves. The English called them "French letters." The French returned the favor by calling them "English raincoats." The origin of the word "condom" is unclear. One story, usually discounted as apocryphal, is that a Dr. Condom (or Condon) advised Charles II (1630–1685), king of England, to use a sheep's intestine for protection. Another story is that they were named for Condom, a town in France. The most likely origin is from the Latin word *condon*, meaning a receptacle, or *cumdum*, meaning a scabbard for a sword. Regardless of its origin, the first known use of the word "condom" in written English appeared in a 1666 report by the English Birth Rate Commission, which blamed the use of condoms for a declining birth rate in England.

In 1677, Antonie van Leeuwenhoek (1632–1723), Dutch inventor of the microscope (see entry 7), became the first person to see sperm. He thought each sperm contained a complete tiny person. As understanding of the role of sperm in reproduction evolved, the Roman Catholic Church declared the use of condoms to be immoral and forbidden because they prevented conception. This did not stop people from using condoms both for protection against disease and to prevent unwanted pregnancy. The only foolproof alternative was abstinence. Condoms were still made of animal tissue and continued to be expensive as well as unreliable.

The big breakthrough in condom production occurred in 1844 when American inventor Charles Goodyear (1800–1860) took out a patent for vulcanized rubber. The vulcanizing process, in which sulfur and natural rubber are heated together, makes rubber stronger, more pliable, and capable of being molded. The first rubber condom was manufactured in 1855. Early rubber condoms were custom-made; a man had to go to a doctor to be measured for one. They only fit the tip of the penis, and they were expensive. Soon, however, manufacturers realized they could make a full-length condom that fit most men, and sales increased. In 1861, the *New York Times* published its first advertisement for a condom. Although they were advertised as protection against disease, at this time condoms were mainly used to prevent unwanted pregnancies.

Many religious and political leaders opposed the use of condoms. Anthony Comstock (1844–1915), a U.S. postal inspector, founded the New York Society for the Suppression of Vice in order to police public morals. In 1873, Comstock and the Society persuaded the federal government to pass what is known as the Comstock Law. This law made it illegal to use postal mail or any transportation system to send material that was "obscene, lewd, or lascivious," which included everything from human anatomy books to birth control devices. It also prohibited the printing and publication of any materials related to abortion, birth control, and venereal disease. Condoms could no longer be advertised. As a result, cases of venereal disease soared.

Unlike German and French soldiers, American and British soldiers were not given condoms during World War I. Consequently, 400,000 cases of venereal disease occurred among the troops. Despite the Comstock Law remaining in effect, condoms to protect against disease became standard issue in the American military by 1931. The Comstock Law was finally declared unconstitutional in 1965; however, advertising of condoms on television remained banned until 1979, and the first television advertisement for a condom was not broadcast until 1991.

Latex, which is made by suspending rubber in water, was invented in 1920. The Youngs Rubber Company became the first company to manufacture latex condoms. These condoms were easier and safer to produce than rubber condoms. They were also stronger and had a longer shelf life. By 1930, assembly line production of latex condoms had brought the price down so condoms became affordable for the middle and working classes. This price drop almost eliminated rubber and skin condoms.

In 1938, because shoddy, leaky condoms were common, the United States Food and Drug Administration (FDA) declared condoms to be a drug under its jurisdiction. This forced manufacturers to test each condom before it was distributed, and it allowed the FDA to confiscate any defective product. In the first month the law was in effect, the FDA seized 864,000 defective condoms.

With condoms now declared a drug, doctors were permitted to provide them as birth control to married couples. From 1930 to the early 1960s when birth control pills became available, condoms were the most commonly used birth control device despite objections from the Roman Catholic Church and other religious groups and social institutions.

The 1960s ushered in the era of oral contraceptives (see entry 44). Up until this time, most women depended on a man's willingness to use a condom to prevent pregnancy. With the arrival of oral contraceptives, a woman could take control of her own fertility. In addition, the development of a range of new antibiotics during the 1950s meant that venereal diseases such as syphilis and gonorrhea were now treatable. Despite improvements such as lubricated condoms, condom use dropped after oral contraceptives became widely available.

The situation changed dramatically in the early 1980s with the understanding that the HIV virus that causes AIDS is sexually transmitted. For many years, an AIDS diagnosis was invariably fatal. Today, drugs allow people with HIV infections to live long, near-normal lives (see entry 54), but at the beginning of the

AIDS epidemic, the disease was terrifying, especially when it became clear that it was not limited to gay men.

Despite resistance from conservative politicians and religious leaders who supported an abstinence-only policy, the United States and many European countries mounted major advertising campaigns promoting condoms as the best way to prevent sexually active people from contracting HIV/AIDS. Safer sex education was introduced in schools, and condom sales rose. A generation has grown up now thinking of condom use not just as birth control but also as a way to prevent disease, echoing the perception of the early condom users of the 1500s.

In the United States, condoms are sold openly in supermarkets and big-box stores in addition to pharmacies and other places where healthcare products are sold. Some university health centers hand out free condoms to students. Condoms are big business. According to Planned Parenthood, as of 2017, more than 450 million condoms were sold in the United States at an average price of $.45 each. Globally, over 5 billion condoms are sold each year.

FURTHER INFORMATION

Cain, Taryn. "History of Condoms from Animal to Rubber." Wellcome Collection, November 19, 2014. https://wellcomecollection.org/articles/W88vXBlAAOEyzwO_

Collier, Aine. *The Humble Little Condom: A History.* Amherst, NY: Prometheus, 2007.

Crockett, Emily. "Watch the Fascinating, Disgusting, Process of How Victorian Condoms Were Made." *Vox,* November 26, 2016. https://www.vox.com/2015/11/29/9813252/victorian-condoms-sheep-intestine

Kahn, Fahd, Saheel Mukhtar, Ian K. Dickinson, and Seshadri Sriprasad. "The Story of the Condom." *Indian Journal of Urology* 29, no. 1 (January–March 2013): 12–16. https://www.ncbi.nlm.nih.gov/pmc/articles/PMC3649591

5. First Wheelchair, 1595

A drawing of what could be a wheelchair appears in Chinese writings from around 600 CE, but there is no proof that this device was used for mobility-impaired individuals. Some sources think it was used to keep important people from getting dirty on unpaved streets and to show their status in society.

The first documented wheelchair, or invalid chair as it was called at the time, appeared in Europe in 1595 (a few sources say 1588). As King Phillip II of Spain (1527–1598) aged, he developed gout—a painful type of arthritis—in the feet. His trusted servant, Jehan de L'hermite (c. 1540–c. 1610), who had come from the Netherlands to educate the king's son, built a moveable chair for the king. The chair had four legs with a small wheel on each leg, an adjustable backrest, and a leg rest. It had to be pushed by a servant and could only be used on smooth indoor floors.

The first self-propelled wheelchair was made in 1655 by Stephan Farffler (1633–1689) of Nuremberg, Germany. Farffler was a watchmaker, and he was either an amputee or had an injury that left him unable to walk. When he was 22 years old, he used his understanding of gear systems in watches to design the first wheelchair that could be moved independently by the user.

Farffler's wheelchair looked like an elongated tricycle with one wheel in front and two larger wheels behind. In fact, some historians consider it the inspiration for the tricycle. To make the wheelchair move, the user turned a hand crank that engaged a gear that turned the toothed front wheel. Apparently Farffler's wheelchair could be used outdoors, as it was reported that he parked it in front of his church every Sunday to attend services.

HISTORICAL CONTEXT AND IMPACT

Despite television and mystery story depictions of criminals who toss bodies into car trunks, adults who cannot walk are heavy and awkward to move. This difficulty, along with the desire to provide expanded opportunities for the disabled, provided the impetus for improvements and special modifications in wheelchairs. Today wheelchairs come in many variations—attendant-propelled, self-propelled, battery-powered, foldable, ultralightweight, specialized for use on beaches, rough terrain, climbing stairs, and playing more than half a dozen sports.

There are wheelchairs made for people who have only one arm. There are sip-and-puff wheelchairs controlled by air pressure generated by inhaling (sipping) or exhaling (puffing) into a tube—for people who cannot use both arms. There are even wheelchairs under development that can be controlled by eye movements. It has taken many experiments and improvements to make wheelchairs meet such a variety of individual needs.

In the eighteenth and early nineteenth centuries, Bath, England, was famous for the healing powers of its natural springs. The sick and disabled flocked to the city to "take the waters," in other words to soak in and drink from the mineral springs. James Heath, a native of Bath, invented a wheelchair in 1750 with two wheels behind and one in front. Although the chair had to be pushed from behind, a bar in the front allowed the passenger to steer. The chair could be rented to transport people to and from the mineral springs. He called it the Bath Chair after his native city. In 1782, John Dawson, another Bath native, became the major maker and seller of the Bath Chair. He used wicker to make the chairs lighter, and soon his chairs outsold all other wheelchairs in England.

In America in the late 1880s, wheelchairs, called rolling chairs, became a recreational phenomenon in Atlantic City, New Jersey. In 1887, William Hayday modified a wooden three- wheeled wheelchair for use on the boardwalk. Rolling chairs were intended to allow those who could not walk the chance to benefit from the healthful sea breeze. They were the only vehicles permitted on the boardwalk.

To make sure his chairs were not stolen, Hayday insisted that users hire an attendant to push the chair. This requirement created a surprising twist. People without disabilities on holiday in Atlantic City took to hiring a chair for a day. The idea of being pushed around by an attendant felt like an affordable holi- day luxury to middle- and working-class people who could not afford servants at home. Although motorized rolling chairs complete with hired drivers were allowed on the boardwalk starting in 1948, they never completely displaced push chairs.

The first rolling chair boardwalk parade occurred in 1902. The tradition has continued. In 2018, contestants in the Miss America Pageant were paraded down the boardwalk in wicker push chairs. The popularity of rolling chairs reached its peak in the 1920s, but one hundred years later, people could still rent a rolling chair on the Atlantic City boardwalk.

Soldiers injured in the U.S. Civil War (1861–1865) accelerated changes in wheelchair technology. The first four-wheeled chair, with two big wheels in front and two small wheels behind, was developed at this time. Push rims also became more common. Push rims are metal wheels covered in rubber that are slightly smaller than the wheelchair's front wheels. They are mounted outside the chair's front wheels and when pushed on by the chair's occupant, they make the chair move forward or backward.

Moving oneself in a wheelchair was an improvement over being pushed, but many of those injured in the war had lost an arm or had no upper body strength and this limited the scope for self-propulsion. In 1896, John Ward of London built the first powered wheelchair. The power source was a gasoline engine. The chair did not catch on, however, until amputees began returning from World War I (1914–1918). Meanwhile, a number of modifications occurred in push chairs, including improved reclining backs, more comfortable seating, foot and leg rests, and the use of cane webbing to make the hairs lighter and easier to handle.

Disabled veterans of World War I introduce a new aspect to wheelchair use— sports. The first documented wheelchair sporting contest occurred in 1923 at the Royal Star & Garter Home for Disabled Veterans in Surrey, England. It consisted of push-rim wheelchair races, an obstacle course, and a lawn bowls contest. Ath- letic contests were popular among veterans. The first team event was a basketball game played in Connecticut in 1945. After that, wheelchair sports took off, and with them came highly modified and often expensive wheelchairs that allowed athletes to play basketball, archery, and tennis and participate in racing, dancing, and other physically competitive activities. The 2005 documentary film *Murder- ball* features a full-contact rugby competition between American and Canadian wheelchair athletes.

Advances also occurred outside the sporting world. Herbert Everest was injured in a mining accident and needed to use a wheelchair. Dissatisfied with

Early wheelchairs had to be pushed by an attendant. Today a variety of self-propelled chairs allow independent travel. (Illustration 163324174 © Patrick Guenette/ Dreamstime.com)

the models that were available, he and Harry Jennings designed the first lightweight collapsible wheelchair in 1932 and went on to establish the company Everest & Jennings to manufacture their creation. The company was still in business as of 2023 and manufactures a full line of wheelchairs and accessories.

Returning disabled World War II veterans wanted greater independent mobility. Canadian inventor George Kline, working with the office of Canadian Veteran Affairs and other veterans' organizations, created the first practical electric wheelchair. It was sold by Everest & Jennings starting in 1956. This was followed by the first electric four-wheeled scooter in 1968. Al Thieme, a Michigan businessman, worked with an engineer to design an electric scooter for a relative who had lost the ability to walk. The demand for the scooter proved so great that Thieme closed his heating and plumbing business and founded Amigo Mobility International. Today the company makes a range of power scooters for personal, hospital, and retail use.

The federal Americans with Disabilities Act was passed in 1990 so that people with disabilities can have the same rights to access and opportunities as anyone else. As a result of this law, businesses, offices, parks, and other public places are required to eliminate barriers that would prevent their use by those with disabilities. For wheelchair users, this has resulted in

modifications such as ramps instead of or in addition to stairs, curb cuts, wheelchair accessible toilet stalls in public places, and placement of elevator buttons, ATMs, water fountains, and similar devices at a level where they can be reached from a wheelchair. Although accessibility is not perfect, between this law and the improvements in wheelchair technology, individuals who cannot walk are acquiring increased ways of moving through the world independently. Since 2008, March 1 has been recognized as International Wheelchair Day.

FURTHER INFORMATION

Kamenetz, Herman L. *The Wheelchair Book: Mobility for the Disabled.* Springfield, IL: Charles C. Thomas, 1969.
Nias, Kay. "History of the Wheelchair." The Science Museum (London), March 1, 2019. https://blog.sciencemuseum.org.uk/history-of-the-wheelchair
Slawson, Mavis C. "Wheelchairs Through the Ages." National Museum of Civil War Medicine, July 9, 2019. https://www.civilwarmed.org/surgeons-call/wheelchairs

6. First American Dentist, 1630

William Dinly (d. 1638), generally recognized as the first "dentist" in America, was not what we would consider a dentist today. In fact, the word "dentist" was not used until the 1750s. Dinly's patients would have called him a barber-surgeon or a tooth drawer. Barber-surgery developed as a profession in the Middle Ages. The combination of barber and surgeon may seem odd today, but barbers were trained to handle sharp blades for shaving, but also needed for bloodletting and amputations. In keeping with the standards of the day, Dinly would likely have performed bloodletting, amputations, and purges, in addition to extracting teeth, all without any anesthetic.

William Dinly sailed from England and arrived in the Massachusetts Bay Colony in 1630 along with two other barber-surgeons whose names have been lost to history. Dinly found his way into the historical record only because he got into trouble with leaders of the rigid Puritan community in Boston who objected to his nonconforming religious beliefs. Despite his heretical views, Dinly was able to practice his trade in Boston and the surrounding area.

During a snowstorm in the winter of 1638, a man in Roxbury, a town outside Boston, was suffering from a toothache. He sent his maid through the snowstorm to fetch Dinly to extract the tooth. On the way to Roxbury, Dinly and the maid were overcome by the storm. Their frozen bodies were found a few days later.

HISTORICAL CONTEXT AND IMPACT

Treating tooth decay has a long history. Skeletons found in northwest India show attempts were made to drill out tooth decay as early as 7000 BCE. People from ancient Greece recorded recipes for treating toothache and dental infections, as did Islamic scholars in the Middle Ages. There was no concept of preventative dentistry. People ignored their teeth until they became painful, and then most tooth problems were solved by pulling the painful tooth.

The era of modern dentistry began in France in 1728 with the publication of a book by physician Pierre Fauchard (1678–1761) that summarized what was known about mouth anatomy, tooth and gum disease, and the methods of treating and restoring teeth. At this time, dentists believed one of two theories about tooth decay—either that decay was caused by an imbalance in the body humors (the fluids blood, yellow bile, black bile, and phlegm) or that decay was caused by food left on the teeth. The food theory was not completely accepted until the mid-1800s. Colgate manufactured the first mass-produced toothpaste in 1873. Before that, dentists made their own tooth-cleaning concoctions of liquid or paste and sold them in bottles or pots (no toothpaste tubes until the 1890s) to patients, who may or may not have seen benefit in using them.

It took more than a hundred years for the first medically trained dentist to arrive in the United States. John Baker, trained in Europe, arrived in Boston in 1763 after practicing dentistry in several European countries. While in Boston, he taught Paul Revere (1734–1818) to make and repair dentures. As a silversmith, Revere would have been familiar with techniques to make the delicate wires used in dentures.

Baker traveled widely in the colonies and at one point is believed to have made dentures for George Washington. These would likely have been made of bone or ivory, not of wood as myth suggests. Baker's status was certainly higher than that of a barber-surgeon. While he was president, Washington invited Baker to dinner, apparently in appreciation of his dentures.

Improvements in the field of tooth and mouth care developed slowly in the United States. Dentistry was considered a trade learned through an apprenticeship rather than through formal education. There were no licensing requirements or government regulations. Almost anyone could call himself a dentist.

Slowly, dental education began to change from a trade learned by apprenticeship to a profession requiring formal education. In 1828, Dr. John Harris established the first dental school in his home in Bainbridge, Ohio. This was a for-profit school with a short course of study. The school lasted only two years before it closed.

The first public institution devoted solely to dentistry was the Baltimore College of Dental Surgery in Baltimore, Maryland. It graduated its first class in 1841, awarding a DDS (Doctor of Dental Surgery) degree after only five months of instruction. The school was absorbed into the University of Maryland in 1923. Maryland was also the first state to examine and license dentists.

The American Dental Association was formed in 1859 and actively promoted formal dental education. Between 1865 and 1870, five new dental schools were established, including one at Harvard that became the first university-affiliated dental college. The Ohio College of Dental Surgery graduated the first medically trained woman dentist, Lucy Hobbs Taylor, in 1866. Meanwhile, Harvard dental school's first class enrolled two Black men, Robert Tanner Freeman and George Franklin Grant. Upon graduating in 1869, they became the first medically trained Black dentists in America.

Despite the push for formal dental education, change was slow, and the dental profession remained dominated by white men. By 1870, only 15% of dentists had graduated from a dental school or dental college. The remainder had either learned dentistry through an apprenticeship or simply proclaimed themselves dentists. Pulling problem teeth remained the procedure of choice for most dentists.

Rapid advances in medicine and hygiene in the early twentieth century affected the way dental health was understood. Slowly, the profession began to shift from reactive dentistry to preventative dentistry. In 1913, the Fones School of Dental Hygiene was established in Bridgeport, Connecticut. An earlier school for hygienists had lasted only one year before being run out of business by dentists who saw hygienists as potential competition. To get around this conflict, the Fones School trained hygienists specifically to provide public school–based dental education and services. From this point forward, dental practices became more specialized and more closely regulated with a greater emphasis on preventative dentistry and preserving teeth rather than on simply pulling them out when they became painful.

Preventative dentistry took a leap forward when carefully controlled studies showed that fluoride could reduce cavities. Grand Rapids, Michigan, became the first city in the nation to add fluoride to its public water supply (see entry 37). As a result, cavities in children decreased by 60%. In 1951, fluoridation became the official policy of the United States Public Health Agency. But fluoridating water was, and to some degree remains, controversial. It generated many lawsuits challenging its safety and the right of the government to "force mass medication" on its citizens. Despite death threats to some of the federal employees in the fluoridation project, the government prevailed. Today more than 200 million Americans receive fluoridated water from their taps. Fluoride is found in toothpaste and mouthwashes. People using wells or other non-fluoridated water sources are encouraged to give their children supplemental prescription fluoride.

In the years since William Dinly lived and practiced as a tooth puller in Boston, dental health has gone from an overlooked aspect of well-being to playing a major role in health and the prevention of many diseases. Dentists are no longer tradesmen but are now fully licensed, highly trained, specialized medical practitioners. To enter dental school in the United States, a student must have an undergraduate degree. Dental school lasts four years and involves both classroom lessons and clinical practice followed by licensing examinations. Many dentists spend an additional one to three years studying a dental subspecialty such as forensic odontology (the collection and evaluation of dental evidence used in law), oral surgery, pediatric dentistry, or sports dentistry. To keep current with advances and maintain their licenses, modern dentists must regularly earn continuing education credits.

FURTHER INFORMATION

American Dental Association. "History of Dentistry." Undated. https://www.ada.org/en/member-center/ada-library/dental-history

Harris, Ruth Roy. *Dental Science in a New Age: A History of the National Institute of Dental Research.* Rockville, MD: Montrose, 1989.

Truman, James S. *The International Dental Journal.* Vol. 24. Philadelphia: International Dental Publication Company, 1903.

Wynbrandt, James. *The Excruciating History of Dentistry.* New York: St. Martin's, 1998.

7. First Microbiologist, 1674

The unseen world has always challenged human imagination and invited investigation. Antonie van Leeuwenhoek (1632–1723) of Delft, the Netherlands, was the first person to see and describe bacteria and protozoa, making him the first true microbiologist. Leeuwenhoek began his adult life as a draper selling cloth and items needed for sewing. His interest in lenses developed because he wanted to be able to closely examine the tightness of the weave of cloth he was buying for his shop. When he was 28 years old, Leeuwenhoek obtained a government position that paid a guaranteed income. This freed him financially to spend much of the next 63 years making lenses, exploring the unseen world, and communicating what he saw to the Royal Society in London.

Around 1665 in London, Robert Hooke (1635–1703), a brilliant man of multiple talents, had used a microscope with a magnifying power of about 30× made by Christopher Cock to examine some biological specimens and fossils. These included the earliest observed microorganism, a microfungus. Hooke published his observations, along with various astronomical observations, in the 1665 book *Micrographia*, but his interest soon shifted to physics, mathematics, and architecture.

Although Leeuwenhoek had likely read Hooke's book during a visit to London, the lenses he made were nothing like those used in the Cock microscope. Over his lifetime, Leeuwenhoek made more than 500 lenses that ranged in size from .08 inches to 2.9 inches (2–73 mm). The smaller the lens, the higher the magnification. He kept his lens-making process secret, something that initially hurt his credibility. Science historians now believe that he made the lenses by pulling hot glass into fine threads and then reheating the ends to make perfect tiny spheres that became the basis for his lenses.

In Leeuwenhoek's microscope, the lens was mounted between two metal plates. The specimen was held in place with a pin, and there apparently was a source of indirect illumination. The apparatus had to be held close to the eye. (A picture of this microscope can be seen in the YouTube video referenced after this entry.)

Leeuwenhoek was educated as a tradesman, not a scientist, but he was precise, organized, and he replicated his work. Most importantly, he asked creative questions with an open mind. In 1674, in a letter to the Royal Society, which was a leading scientific organization in Europe, he described the presence of thousands of tiny eel-like creatures in a drop of lake water. In a 1677 letter, he described a teeming world of microscopic animals in a drop of rain water that had stood in a new, clean earthen pot for several days. This made him the first person to describe protozoa and bacteria, and his careful lifelong study of these made him the first microbiologist.

Leeuwenhoek's descriptions were met with skepticism at the Royal Society. Single-cell organisms invisible to the naked eye had never before been described. To combat the skepticism, Leeuwenhoek insisted that the Royal Society send trustworthy men to Delft to see his discoveries for themselves. A committee of nine, including three ministers and members of the Royal Society, eventually looked through Leeuwenhoek's microscope and confirmed his descriptions.

Leeuwenhoek later observed the sperm of men and different animals, studied insect reproduction, described vacuoles in cells, and witnessed blood cells flowing through capillaries. He was also the first person to use a stain to color his specimens to make them more visible. His stain was made of saffron, which produced a yellow-orange color.

Leeuwenhoek's discoveries made him famous during his lifetime. By the time he died at age 91, he had written more than 550 letters to the Royal Society documenting his discoveries. He willed his lenses and specimens to the Royal Society. Modern scientists have found that the lenses had magnification powers ranging from 50× to 300×, far greater than any other lenses known to exist at that time.

HISTORICAL CONTEXT AND IMPACT

As early as 165 BCE, the Chinese created a crude magnifying scope using some sort of a lens mounted in a tube containing water. The amount of water in the tube determined the degree of magnification, but it also created distortion.

Lenses that could be used in spectacles (eyeglasses) were developed in Italy in the late thirteenth century. Two men from Tuscany both claimed on their gravestones that they had invented spectacles. Salvano d'Aramento degli Amati (d. 1284) kept his process secret. Allessandro della Spina (d. 1317) supposedly shared his invention with others. Little else is known about these men. However, by the early 1300s, references to spectacles could be found in various writings in Europe.

Dutch lens-maker Zacharias Janssen (b. 1585) and his father Hans are thought to have made the first compound microscope, although this claim was made after their deaths and cannot be verified. The Janssen microscope used multiple lenses aligned in a metal tube. This provided significantly better magnification than a single lens. The additional lenses, however, also increased distortion of the object being viewed.

By the 1660s, improved Janssen-style microscopes were being made in other parts of Europe. Christopher Cock of London made one for Robert Hooke around 1665. The actual microscope made by Cock and used by Hooke can be seen in the Billings Microscope Collection of the National Museum of Health and Medicine in Silver Spring, Maryland.

Leeuwenhoek's revelations about a world seen only under the microscope forced a reconsideration of how life is created. At the time he began writing letters to the Royal Society, there was a general belief in the spontaneous generation of living organisms from nonliving matter. Although it was understood that mating between male and female animals produced offspring of the same species, this concept was not applied to animals such as tiny insects for which mating had not previously been observed. For example, sand flies were thought to arise from sand and maggots from meat, as if God or a divine spirit in the air spontaneously breathed life into the inanimate. This belief probably explains why ministers from three different religious denominations were included in the committee that investigated Leeuwenhoek's observations.

Leeuwenhoek's discovery of the microscopic world inside a drop of water challenged the theory of spontaneous generation, but it took until the 1860s for Louis Pasteur (1822–1895) to definitively disprove spontaneous generation by performing a series of experiments to show that microorganisms exist and can reproduce to cause fermentation, food spoilage, and disease.

In 1890, German bacteriologist Robert Koch (1843–1910) published what are known as Koch's postulates, another advance in understanding bacteria and their role in disease that would not have occurred without Leeuwenhoek's discovery of microbes. In summary, Koch's postulates state that the organism causing a disease is found in all individuals who have the disease. The organism can be isolated and

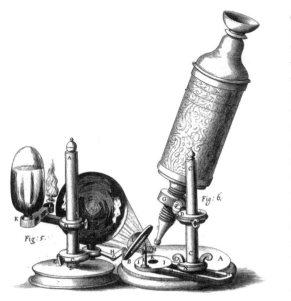

Early microscope similar to the one the first microbiologist Antonie van Leeuwenhoek used. Indirect
light is provided by the candle and reflected by the
mirror. (Wellcome Library, London)

cultured. When introduced into a healthy individual, the cultured organism will cause the disease, and when re-isolated from that individual, the organism will be identical to the one that caused the initial disease. These fundamental truths eventually resulted in advances in pharmaceutical medicine, such as the development and testing of antibiotics in the 1940s and, more recently, the development of antiviral drugs.

Microscopes, too, have undergone improvements that have increased magnification and reduced distortion. A giant advance occurred with the development of the electron microscope by Max Knoll (1897–1969) and Ernst Ruska (1906–1988). This microscope uses an accelerated beam of electrons that are shorter than visible light and thus can achieve a higher resolution and magnifications up to 10,000,000×, while the best light microscopes have a resolution of 2,000×. The drawback to the electron microscope is that the sample is killed by the electron beam, so it is impossible to see movement or actions such as blood flowing through capillaries. Ernst Ruska shared the Nobel Prize in Physics in 1986 with Gerd Binnig (1947–) and Heinrich Rohrer (1933–2013) for advances in optics and microscopy.

Today, microbiologists tend to specialize in certain types of microorganisms, including bacteria, viruses, algae, fungi, and some types of parasites. According to the Bureau of Labor Statistics, about 19,400 microbiologists were employed in the United States in 2021. These microbiologists work in areas such as water and food safety that affect everyone on a daily basis.

Individuals with two-year associate's degrees can work as microbiology laboratory technicians in industrial, medical, and public health fields. With a four-year bachelor's degree, individuals are employed as microbiologists in the fields of food science, water purification, wastewater treatment, environmental science, human and veterinary medicine, and public health. With a master's degree in microbiology individuals can supervise laboratories, while those with a PhD do research, teach at the university level, or work for the government in public policy. In 2021, the Bureau of Labor Statistics estimated the annual salary for microbiologists as ranging from $48,000 for technicians with a two-year degree to more than $137,000 for PhDs.

FURTHER INFORMATION

"History of Microscopes." Microscope, Undated. https://www.microscope.com/education-center/microscopes-101/history-of-microscopes

Lane, Nick. "The Unseen World: Reflections on Leeuwenhoek (1677) 'Concerning Little Animals.'" *Philosophical Transactions of the Royal Society* 370, no. 1666 (April 19, 2015). https://royalsocietypublishing.org/doi/10.1098/rstb.2014.0344

LSG Science. "History of the Microscope." YouTube, October 2, 2016. https://www.youtube.com/watch?v=TjyOQmd

8. First Clinical Trial, 1747

A clinical trial is defined by the World Health Organization as "any research study that prospectively assigns human participants or groups of humans to one or more health-related interventions to evaluate the effects on health outcomes." By this definition, James Lind (1716–1794), a Scottish naval doctor, performed the first clinical trial.

Lind was born into a family of Edinburgh merchants and received his medical training through an apprenticeship, which was the common way of training doctors at that time. In 1739, he joined the Navy as a surgeon's mate. By 1747, he had been promoted to surgeon on the HMS *Salisbury*. He performed the first clinical trial for curing scurvy on this ship.

Scurvy is a disease caused by chronic vitamin C deficiency. The body cannot store vitamin C. The vitamin is water soluble and excess amounts are soon removed in urine. To remain healthy, people need a regular supply of the vitamin. The best sources of vitamin C are citrus fruits—oranges, lemons, limes, and grapefruit. At the time Lind was in the Navy, these fruits or their juices were not provided to British sailors. Interestingly, the Spanish knew their sailors needed citrus fruits to remain healthy, which is why Christopher Columbus, on his second voyage to the New World in 1493, brought orange seeds from Spain to plant on Hispaniola. When the trees matured, future sailors would have a supply of fresh fruit for the return voyage to Spain. Somehow, possibly because citrus fruits are not native to the British Isles, the British Navy failed to recognize the cause of scurvy or learn how to prevent it until Lind performed his clinical trial.

Scurvy can be fatal. Visible symptoms include fatigue; swollen, spongy, and purplish gums that are prone to bleeding; loose teeth; bulging eyes; dry, cracked skin that bruises and bleeds easily; and swollen, painful joints. Metabolic symptoms include suppression of the immune system; failure to absorb iron; disruption of cholesterol metabolism; and a decrease in the synthesis of neurotransmitters, the chemicals that facilitate nerve impulse transmission. In Lind's time, up to half the sailors on a long voyage on British ships would die from scurvy.

When sailors on the *Salisbury* developed scurvy symptoms, Lind conceived his trial. He chose 12 men with very similar scurvy symptoms, isolated them, and assigned a healthy sailor to care for them. Lind's hypothesis, which turned out to be wrong, was that acidic substances could cure scurvy. During the trial, the men continued to eat their regular diet of water, gruel sweetened with sugar in the morning, mutton broth or a light pudding at mid-day, and an evening meal of barley or rice, raisins, a tapioca-like starch, and wine.

Lind divided his trial subjects into pairs. Each pair received a different supplement to their regular diet. The supplements were a quart of cider daily, 25 drops of elixir vitriol (a dilute sulfuric acid) three times daily, two spoonfuls of vinegar three times daily, seawater, two oranges and one lemon daily, or a honey-sweetened remedy recommended by a hospital surgeon.

After six days of treatment, one man who had eaten the oranges and lemons had recovered to the point where he could return to duty. The other man on this diet recovered fully a few days later. None of the men on the other supplements improved during the two-week trial.

Even though he was wrong in his hypothesis about the effectiveness of acidic treatments, Lind recognized that oranges and lemons had curative properties. Citrus fruits had to be imported into Britain. This made them expensive, so Lind was reluctant to recommend the fruits as an addition to sailors' diets. It took 50 years before the British Navy made lemon juice mandatory in sailors' diets. Soon after that, lemon juice was replaced with lime juice, even though lemon juice contains four times more vitamin C than lime juice. This is the reason why, beginning in the 1850s, sailors in the British Royal Navy were called limeys.

HISTORICAL CONTEXT AND IMPACT

The idea of testing treatments in a structured way, with individuals assigned at random to different treatments or as controls, developed slowly. In 1863, Austin Flint (1812–1886) introduced the concept using a placebo—a substance that has no therapeutic effect in a control group—to offset any psychological effects of treatment. Flint, an American cardiologist, divided his patients hospitalized with rheumatic fever into groups and treated the swollen joints of each group with liniments containing different substances: opium, camphorated oil, aconite (a poison better known as wolfbane or monkshood), and an extremely dilute tincture derived from soaking wood in water. This final group was his control group. All the groups were put on bed rest. Flint found that the control group improved at the same rate as the treated groups. In other words, symptoms of rheumatic fever gradually diminished on their own with just bed rest.

The next big advance in clinical trials was the concept of double blinding. In a double-blind trial, neither the patient nor the doctor knows whether the patient is given a drug or a placebo. This eliminates unconscious psychological biases such as the doctor assigning sicker patients to the treatment group and making less sick individuals the controls.

The first double-blind trial occurred in England in 1943 in a test of a drug against the common cold. A nurse assigned alternate patients either a placebo or patulin, an extract of a penicillin fungus that grows on stored apples. Neither the doctor nor the patient knew what they had received. Although the drug proved useless, the trial was important as the first rigorously controlled double-blinded study.

Another improvement in clinical trials occurred in 1946. Sir Austin Bradford Hill (1897–1991), a British epidemiologist and statistician, organized a trial of the antibiotic streptomycin against tuberculosis. Streptomycin was a new drug. There was not enough of it available to treat every tuberculosis patient, so it was not considered unethical to withhold treatment from some people and view them as a control group. Patients were not told that they were part of a trial, something that would be illegal today.

In the study, each patient was given a number, and using a random number table they were assigned either to receive the drug and bed rest or to receive bed rest and a placebo. Neither the trial coordinators, the doctors nor the patients knew whether they were receiving trial treatment or the placebo. This was the first rigorous double-blinded and randomized clinical trial.

Today the gold standard for clinical trials remains the double-blinded randomized trial. In the United States, clinical trials proceed in three stages and usually take many years. Phase I begins after animal testing. Investigators file an Investigational New Drug application and receive permission from the Food and Drug Administration (FDA) to begin the trial. Phase I usually involves only between 30 and 70 volunteers. The goal of this phase is to determine if the drug is safe and if it produces the expected response.

If the drug moves to Phase II testing, at least several hundred volunteers are involved. The goal of Phase II is to determine the correct dosage of the drug and to study its safety in a larger population similar to the one that will be receiving the drug. This phase usually takes several years.

If a drug moves to Phase III, thousands of volunteers are enrolled in a randomized double-blind study. The numbers need to be large to determine statistical differences in effectiveness and frequency of side effects between the experimental group and the control group.

All aspects of clinical trials are regulated and overseen by the FDA. If the FDA approves the drug after reviewing Phase III data, physicians can then prescribe the drug. Even after a drug is approved, the manufacturer may be required to do additional studies and review side effects in select subpopulations such as people with special health conditions. Post-licensure evaluations are sometimes called Phase IV.

An ethical and regulatory framework for clinical trials has evolved along with trial design. The 1947 Nuremberg Code outlined 10 basic criteria for human medical research, including voluntary participation and informed consent. Since then, more protections have been added, including the right to withdraw from a trial for any reason. Today, research proposals involving humans in medical and university settings must be approved by an independent institutional review board that determines whether the trial is generally safe and if the benefits of the research outweigh the risks. May 20, the day in 1747 that James Lind began his scurvy trial, is now recognized as International Clinical Trials Day.

FURTHER INFORMATION

Bhatt, Arun. "Evolution of Clinical Research: A History Before and Beyond James Lind." *Perspectives in Clinical Research* 1, no. 1 (January–March 2010): 6–10. https://www.ncbi .nlm.nih.gov/pmc/articles/PMC3149409

Kaptchuk, Ted J. "A Brief History of the Evolution of Methods to Control Observer Biases in Tests of Treatments." *JLL Bulletin*, 2011. https://www.jameslindlibrary.org/articles/a -brief-history-of-the-evolution-of-methods-to-control-of-observer-biases-in-tests-of -treatments

Milne, Iain. "Who Was James Lind, and What Exactly Did He Achieve?" *JLL Bulletin*, 2003. https://www.jameslindlibrary.org/articles/who-was-james-lind-and-what -exactly-did-he-achieve

Nellhaus, Emma M., and Todd H. Davies. "Evolution of Clinical Trials throughout History." *Marshall Journal of Medicine* 3, no. 1 (2017): 41. https://mds.marshall.edu/mjm /vol3/iss1/9

9. First Female Doctor with a Full Medical Degree, 1754

After years of struggle against an all-male medical system, Dorothea Leporin Erxleben (1715–1762) became the first female to receive a university medical degree. Dorothea's father, a progressive middle-class physician in Quedlinburg, Prussia (now part of Germany), believed that men and women should be educated equally. This was an unusual—almost radical—position at the time, but it resulted in Dorothea receiving the same education as her brother. Together, Dorothea and her brother Christian studied French, Latin, and theology with a local pastor and theoretical and practical medicine with their father. Dorothea proved to be an intellectually gifted student.

Both siblings planned to enter the University of Halle, which today after several mergers is called the Martin Luther University of Halle-Wittenberg. Before they could enroll, Christian was called up for military service. At that time, it was unthinkable for an unchaperoned woman to attend university, so Dorothea's plans were delayed.

In 1740, after completing his military service, Christian applied to enter the university and was accepted. Dorothea was denied admission. Her solution was to petition King Frederick II, later known as Frederick the Great, for permission to attend. Frederick had progressive leanings, and eventually permission was granted.

Dorothea's acceptance into the University of Halle created controversy. The general attitude at the time was that higher education for women was unnecessary. One critic argued that the law did not allow women to practice medicine, so there was no point in a female earning a medical degree and taking a spot that should go to a man. In response, and with her father's support, Dorothea published *A Thorough Inquiry into the Causes Preventing the Female Sex from Studying*. In the book, she argued that women were an untapped source of talent and experience. Educating them would benefit the country. Her father wrote an introduction to the book, advocating for changes in the university system to accommodate the education of women.

Despite her acceptance into the university, in 1742 at the age of 26 (an old maid by the standards of the time), Dorothea chose to delay her medical education and married Johann Erxleben, a Lutheran deacon and widower with five children. Over the next few years, she also had two daughters and two sons. Her father died in 1747 leaving his family in debt, and soon afterward, her husband became seriously ill. It appeared that a university medical degree was beyond her reach. To support her large family, Dorothea Erxleben began practicing medicine without a degree.

Although it appears that Erxleben was a skilled and respected medical practitioner, local physicians resented her success. In 1753, three doctors filed a legal suit against her claiming that she was a medical fraud who practiced medical quackery. They also claimed that she illegally called herself a doctor, visited patients, and accepted money for her treatments. In response, she claimed that she never called herself a doctor and asked the suing doctors to provide proof. She agreed that she visited patients and sometimes accepted payment for her services, but she also stated that she treated poor people for free, something doctors generally did not do. Then she challenged the three doctors suing her to let her take the university medical examination and to take it themselves and see how everyone scored.

This challenge infuriated the suing doctors. They refused to take the examination. Instead, they accused her of being a witch. They also claimed that females did not have the intelligence to be physicians and that her pregnancies, one of which corresponded to the time of this dispute, prevented her from successfully practicing medicine. They even charged her with malpractice when one of her patients died.

Erxleben's response was to bring the matter before King Frederick II who had initially approved her request to study medicine. The king's solution was to permit Erxleben to take the university medical examination and to submit a required dissertation. If the results proved to be of high quality, then she would be granted a medical degree. Fortunately, the rector (most senior official) of the university was a believer in education for women and supported this decision.

Erxleben impressively passed all the exams, which were given in Latin, and submitted a dissertation in which she argued that many doctors undertook unnecessary cures either because they intervened too quickly in minor illnesses, or because they gave in to the urging of their patients for a quick, stronger-than-necessary cure. The examination results and dissertation were accepted, and Erxleben became the first woman to be awarded a university medical degree on June 12, 1754. With this degree, she gained the legal right to call herself a doctor and to practice medicine as a recognized physician.

After gaining her degree, Erxleben continued to treat patients until her death in 1762. However, no other woman graduated with a medical degree from Erxleben's university (now known as Martin Luther University Halle-Wittenberg) until 1901. In 1987, Germany recognized Dorothea Erxleben's achievement as the first woman to earn a university medical degree by issuing a postage stamp with her picture.

HISTORICAL CONTEXT AND IMPACT

In the ancient world, women were healers. Women gathered plants, prepared food, and tended the sick. Over time, they became herbalists with a working knowledge of which plants could reduce fever, calm a cough, or fight infection. Add to this the prevailing belief that a deity could directly influence the health of an individual and it is not surprising that many female healers were associated with religious institutions.

In ancient Egypt, the goddess Isis had, among many attributes, the ability to heal ordinary people. Priestesses who lived in temples dedicated to Isis are thought to have practiced as early healer-physicians. Other women in ancient Egypt appear to have been trained to deliver babies, making them early midwives. In ancient Greece, the goddess Hygeia was associated with preventing illness and the goddess Panacea with curing disease. In temples dedicated to these goddesses, both priests and priestesses prescribed cures based on the interpretation of oracles. Later, women in both Greece and Rome openly practiced medicine in competition with men.

After the fall of the Roman Empire, much scientific and medical information was lost or destroyed. Convents and monasteries became centers for the information that was preserved, and the task of healing fell to both men and women in religious orders, once again combining religious faith and medicine. Nevertheless, by the 1300s, the role of women in the church had been substantially diminished as had their role as healers. Outside the church, laws made it impossible for women to openly practice medicine or deliver babies. Women who continued to do these things risked being accused of witchcraft and persecuted.

The Renaissance brought new ideas about art, architecture, and science, but women were generally believed to be biologically incapable of meaningful participation in these fields. A few enlightened thinkers emerged to counter this idea, but among the common people, attitudes changed slowly. For example, in the 1750s, Dorothea Erxleben was accused of being a witch for practicing healing. Women were allowed to nurse the sick and aid in childbirth, but only under the supervision of a male physician.

Things were only slightly better in America a century later. Elizabeth Blackwell (1821–1910), a British woman who immigrated to the United States as a child, was rejected by almost a dozen medical schools because of her sex. When she applied to Geneva Medical College in rural New York, the students were asked whether the college should admit her. The students assumed that her application was a prank and agreed to the college sending her an acceptance letter, never believing she would attend. Not only did she attend but also in 1849, after facing discrimination and scorn from male students and professors, she became the first woman to earn a medical degree from an American college. Even with a medical degree, Blackwell had few patients and could not gain hospital admitting privileges or work at a clinic. Her solution was to form the New York Infirmary for Women and Children in partnership with her sister, who was also a doctor. They were joined by another female physician.

Since the days of Elizabeth Blackwell, the situation for women in the United States who want to practice medicine has changed considerably. According to the Association of American Colleges, in 2020, women made up 51.9% of medical students. Graduate female doctors dominate the fields of obstetrics and gynecology, pediatrics, and allergy and immunology while men still dominate orthopedics, neurological surgery, and interventional radiology. In other medical specialties, the gender balance is approximately equal. In most other industrialized nations, men continue to dominate the field of medicine.

FURTHER INFORMATION

Ehrenreich, Barbara, and Deirdre English. *Witches, Midwives, and Nurses: A History of Women Healers*. Detroit: Black & Red, 1973.

Jefferson, Laura, Karen Bloor, and Alan Maynard. "Women in Medicine: Historical Perspectives and Recent Trends." *British Medical Bulletin* 114, no. 1 (June 2015): 5–15. https://academic.oup.com/bmb/article/114/1/5/246075

Wynne, Rhoda. "Saints and Sinners: Women and the Practice of Medicine Through the Ages." *JAMA* 283, no. 5 (February 2, 2000): 668–669. https://jamanetwork.com/journals/jama/fullarticle/1843228

10. First Publicly Funded American Psychiatric Hospital, 1773

The first publicly funded psychiatric hospital was founded in Williamsburg, in the colony of Virginia, on October 12, 1773. At its founding, its official name was the Publick Hospital for Persons of Insane and Disordered Minds. The local population simply called it the Mad House. In 1841, the hospital changed its name to Eastern Lunatic Asylum and in 1884 to Eastern State Hospital. It still operates under that name today.

The impetus for founding a publicly funded institution to care for the mentally ill came from Francis Fauquier (1703–1768), the governor of Virginia from 1758 to 1768. Fauquier was the son of a doctor and appears to have been familiar with psychiatric institutions in England. In 1766, he requested the General Assembly, the legislative branch of the colony, to authorize the building of a hospital for individuals who were deemed by society as deprived of their reason and unable to help themselves. A resolution was passed, but nothing else happened, so Fauquier brought up the subject the following year. He died before any concrete action could be taken. Nevertheless, Virginia's next governor, Norborne Berkeley, 4th Baron Botetourt (1718–1770), known in Virginia as Lord Botetourt, continued to press for a mental health facility, and in 1771, construction of a building on land owned by the College of William and Mary began under the supervision of well-known builder Benjamin Powell (1730–1791).

The original institution was a cross between a jail and a hospital. The brick building consisted of meeting rooms, living quarters for a supervisor and his family, and cells that could accommodate 24 patients. Outside, fenced courtyards allowed the patients to exercise outdoors. This feature was quite advanced for the time, as many similar institutions in Europe simply locked patients in their cells.

The first patient was admitted to the facility on October 12, 1773. It took about 30 years before all 24 cells were filled, partly because the objective was to treat patients and then return them to society once their mental health improved.

The first staff consisted of Dr. John Galt (1744–1808), a physician who had received his medical training in Scotland; his wife, who served as hospital matron and cared for female patients; and a few enslaved persons who performed daily maintenance chores.

In 1841, supervisor John Minson Galt II (1819–1862), grandson of the first superintendent, changed the hospital's name to Eastern Lunatic Asylum. Calling it an asylum rather than a hospital was intended to indicate that it was a place for people who needed to live in a sheltered environment rather than for those who needed to be locked up in cells.

Galt II made major, and mainly positive, changes in the asylum that were far ahead of most mental health facilities at the time. He advocated a system called "moral management" for the mentally ill that involved allowing recovering patients to go into Williamsburg during the day and mingle with the townspeople, as long as they exercised good self-control. Galt also established a shoemaking shop, carpentry shop, a sewing room, library, and a game room for the residents. During Galt II's time as superintendent, the hospital housed between 200 and 300 patients, referred to as "inmates," in seven buildings. About half the patient population participated in the moral management program.

Over the next 20 years, the hospital added three buildings and about 100 additional patients. A fire destroyed one building in 1876. A second fire in 1885, blamed on faulty electrification of the buildings, destroyed the original building and five newer buildings, displacing 224 patients. During the Civil War, the Union Army captured the hospital and surrounding area. The Army took over the asylum in May 1862 after all but one of the staff had fled. The buildings were later returned to the state of Virginia.

Following the Civil War, the concept that mental illness was inherited and usually incurable became popular, and the philosophy of moral management and integration of the mentally ill into society was abandoned. Instead, the hospital was used for long-term custodial care with little attempt to provide psychiatric treatment. With no effort to return patients to society, the number of patients rapidly increased. By 1935, the hospital housed 2,000—mostly long-term—psychiatric patients. Crowding was a serious problem, and the institution was out of space to expand in its original location.

Beginning in 1937 and extending through 1968, Eastern State Hospital gradually moved to a new location on the outskirts of Williamsburg where it remains today. The original site was returned to the College of William and Mary, which used it for student housing until about 2006. The original 1773 public hospital building was reconstructed as part of Colonial Williamsburg Foundation's living history museum. It opened to the public in June 1985.

HISTORICAL CONTEXT AND IMPACT

Historically, people with mental health disorders were treated within the family, and in wealthy families, they were hidden away in private facilities. However, there were some early exceptions. For example, a hospital in ninth-century Baghdad had a wing for mentally ill patients too violent to be cared for by their families. In America in 1752, before the establishment of the Williamsburg hospital, some prominent Quakers arranged for the Pennsylvania Hospital in Philadelphia to set aside a row of cells for the mentally ill, but psychiatric patients always remained a small part of the hospital population.

Well into the 1800s, society expected families to care for their mentally ill relatives. Mental illness was generally considered shameful and a blot on the family's standing in society, so people with psychiatric disorders were often kept locked up and out of sight. Public institutions were scarce. In Britain, the first publicly funded hospital for the mentally ill was not opened until 1812, in other words, not until 39 years after the opening of the Williamsburg mental hospital.

From the mid-1700s through the early 1900s, commitment to psychiatric facilities was sometimes used to control female family members. At a time when men were the unquestioned heads of the family, a husband or brother could have an unwanted or nonconformist wife (or an unmarried daughter or sister who had embarrassed the family) committed to what was then called an insane asylum on his word alone or with the assistance of a cooperative doctor. Some mentally healthy women remained incarcerated in these institutions for years simply because their behavior did not fit a man's idea of how a woman should think and act.

In the 1850s, activist Dorothea Dix (1802–1887) traveled throughout the United States advocating for better living conditions for institutionalized mentally ill persons. Dix was an advocate of the moral management approach. This philosophy promoted humane living conditions, medical treatment, and the potential for the mentally ill to engage more fully with society. Through her advocacy, Dix persuaded the U.S. government to build 32 publicly funded state psychiatric hospitals. Unfortunately, by the early 1900s, the concepts of moral management had again fallen out of favor, and these large state facilities became understaffed, underfunded warehouses offering little treatment or hope of rejoining society for the patient. By 1935, for example, Eastern State Hospital housed 2,000 patients and was grossly overcrowded.

Beginning in the late 1950s, a better understanding of mental illness and the development of drugs to treat or moderate severe psychiatric symptoms resulted in a gradual movement to deinstitutionalize the mentally ill. Many large specialized psychiatric hospitals closed, especially as the patients' rights movement gained strength. In the 1950s in the United States, there were approximately 560,000 institutionalized mentally ill patients, or by another measure, 339 psychiatric beds for every 100,000 people. By 1980, the number had dropped to 130,000 institutionalized people, and by 2000, there were only 22 psychiatric beds for every 100,000 people. This trend has continued into the present day with more psychiatric patients being treated at outpatient clinics, at acute crisis care centers, through short-term hospital stays, in small residential homes, and by private physicians. Advocates of deinstitutionalization feel these options provide the mentally ill with more choices, a better lifestyle, and greater personal dignity.

Critics of the deinstitutionalization movement claim that many residential homes are understaffed with poorly trained caregivers and that short-term care is often inadequate to effect long-term mental health improvement. They also point out that the mentally ill frequently interact with the legal system and repeatedly end up in prison where they have little or no access to mental health treatment. Some critics point to the large percentage of homeless individuals who have psychiatric problems and question whether many of these people would have a better life living in an institution with regular access to food, shelter, and healthcare professionals.

FURTHER INFORMATION

"Eastern State Hospital W-40-b." Southeastern Virginia Historical Markers, March 22, 2012. https://sevamarkers.umwblogs.org/2012/03/22/eastern-state-hospital-w-40-b

Groopman, Jerome. "The Troubled History of Psychiatry." *New Yorker*, May 20, 2019. https://www.newyorker.com/magazine/2019/05/27/the-troubled-history-of-psychiatry

Siske, James Harding. "A History of the Eastern State Hospital of Virginia under the Galt Family (1773–1862)." *Dissertations, Theses, and Masters Projects.* Paper 1539624478, 1949. https://scholarworks.wm.edu/etd/1539624478

11. First Vaccine, 1796

The first vaccination was performed against smallpox, a virulent, contagious, viral disease that kills about one-third of people who contract it. Those who survive are often left with ugly scars and disabilities. In 1796, British physician Edward Jenner (1749–1823) was the first person to develop a safe, reproducible method of vaccination to prevent infection from smallpox.

Edward Jenner was born in Berkeley, Gloucestershire, the eighth of nine children of a minister. At age 14, he was apprenticed to a surgeon. He completed his medical training in 1770 at St. Georges Hospital in London while working under one of the best surgeons of the time. In addition to studying medicine, Jenner was seriously interested in natural science and geology. He helped to classify many of the specimens that explorer Captain James Cook (1728–1779) brought back from his first voyage to the Pacific. Cook was impressed enough to invite Jenner to sail with him on his second voyage, but Jenner declined. Instead, he returned to Gloucestershire in 1772 to practice medicine.

Gloucestershire was a rural county, and Jenner had heard since childhood folk wisdom that claimed people who contracted cowpox would never develop smallpox. Cowpox is a mild disease of cattle that causes pus-filled blisters on the udder. People who had direct contact with the blisters during milking developed a fever and a rash similar to, but much milder than, the symptoms caused by smallpox. Full recovery from cowpox took about two weeks, and unlike smallpox, the disease was never fatal. However, recovering from cowpox appeared to protect people from contracting smallpox. We now know this happens because the cowpox virus is a close relative of the smallpox virus and so the immune system changes that occur when someone is infected with cowpox will protect them against smallpox.

During the late 1700s, some farmers intentionally exposed family members to cowpox to protect them from smallpox. What set Jenner apart was that he proved the protective qualities of cowpox could not only be obtained by contracting the disease directly from a cow but could also be passed on from person to person.

Jenner performed his first smallpox experiment on May 14, 1796. He found a milkmaid, Sarah Nelmes, who had fresh cowpox sores on her hands, took some of the pus from her sores, and scratched the pus into the arm of James Phipps, the eight-year-old son of his gardener. Phipps became mildly ill with cowpox but recovered within two weeks. About six weeks later, Jenner scratched pus from the blister of a person with smallpox into Phipps's arm. The boy did not become sick even after a second intentional exposure to smallpox.

Jenner submitted his findings to the Royal Society in London, but his paper was rejected. However, in 1798, after vaccinating several other people, he self-published a booklet describing his technique, which he called vaccination, after the Latin *vacca* meaning cow. Not everyone approved of his vaccinations. Members of the clergy objected on the grounds that vaccination interfered with God's plan and that introducing material into a human from a diseased animal was unclean. Some doctors objected because they made money from variolation, another more dangerous technique to protect against smallpox. The popular press ridiculed Jenner, publishing cartoons showing vaccinated people developing cow horns and udders.

Despite opposition to vaccination, Jenner persisted in recruiting doctors to perform the procedure and supplied dried cowpox material free to any doctor who requested it. By 1801, about 6,000 Britons had been vaccinated. As more people were safely vaccinated, the procedure became accepted. Before his death, Jenner was recognized as a medical hero across Europe.

HISTORICAL CONTEXT AND IMPACT

Smallpox is an ancient disease. In Egypt, the mummified body of King Ramses V who died in 1145 BCE suggests that smallpox caused his death. Smallpox has also been described in Chinese writings dating from 1122 BCE and in early Sanskrit writings in India. In the period between 400 CE and 600 CE, the virus was introduced into Europe by traders. The Portuguese carried the disease to Africa. African slaves carried it to the Caribbean, and European colonists introduced it into North America where it wiped out entire tribes of native people.

People in the ancient world did not know that smallpox can be caused by two variants or strains of the virus, *variola major* and *variola minor*. What they did recognize was that some people had relatively mild symptoms (caused by *variola minor*) while others had severe symptoms that could be fatal (caused by *variola major*). Cultures across the globe from China and India to the Middle East and North Africa tried exploiting the difference in the intensity of the disease as a way of controlling it. The process was slightly different in each culture, but the basis for it was the same everywhere.

A sample of pus would be taken from a person who had a mild case of small-pox, and the pus would be dried to weaken the virus further. A person who had not had smallpox was then exposed to the dried pus. In China, the material was ground into a powder, put into a tube, and puffed up the nose of the uninfected person. In Sudan, a cloth with dried pus was wrapped around the arm of an unin-fected child, and in Turkey, dried pus was inserted under the skin with a needle. Regardless of the method, the goal was to cause a mild case of smallpox that would then provide lifetime immunity against the more virulent strain of the disease. In many cases, this process, which came to be called variolation, was suc-cessful. Still, the death rate from variolation was about 3%.

Variolation was introduced into Europe by Lady Mary Wortley Montagu (1689–1762), wife of the British ambassador to Turkey. In Turkey, she observed the variolation process and was impressed enough to persuade the doctor attached to the British embassy to variolate her son. Later, during a smallpox outbreak after she returned to London, she had her daughter variolated and persuaded the queen to allow some of her children to undergo the procedure. Variolation remained the main way of controlling smallpox until Jenner offered a safer alter-native by using cowpox pus.

Despite public skepticism, Jenner's vaccination process quickly spread to America. Jenner sent dried cowpox pus to Dr. Benjamin Waterhouse (1754–1846) in Boston. Initially, Bostonians were skeptical of the procedure, but an outbreak of smallpox in the city changed their minds. Thomas Jefferson was sent a sample of the material and is said to have vaccinated 200 of his slaves and neighbors. As president, he created the National Vaccine Institute, and his suc-cessor, James Madison, signed the National Vaccine Act of 1813 whose purpose was to assure the purity of the smallpox vaccine and to make it available to every American.

Jenner's success stimulated interest in creating vaccines against other dis-eases. Louis Pasteur (1822–1895) made vaccines against anthrax and rabies using a technique that weakened the disease-causing virus to the point where it stimulated an immune response but did not cause illness. Other vaccines were created by using killed virus or modifying toxins that infective bacteria produced. A vaccine against polio in 1955 made Jonas Salk (1914–1995) an international hero, and the following years saw vaccines developed against measles, mumps, and rubella (German measles). Vaccines became safer as advances in genetic engineering made it possible to stimulate immunity by using only part of the virus. In 2020, the COVID-19 vaccine was the first widely used human vaccine to use messenger RNA (mRNA), an instruction protein that causes the body's own cells to synthesize the material that stimu-lates immunity.

Advances in vaccines were not always greeted with universal enthusiasm. Right from the first vaccine for smallpox, there were always some people who rejected vaccination. The modern anti-vaccination movement began in the early 1980s. It remains especially strong in the United States and Britain. Objec-tions to vaccination range from fear that vaccines cause a variety of develop-mental problems—a theory that has been scientifically rejected—to mandatory vaccination for school attendance being seen as an infringement on personal liberty.

The Cow Pock — or — the Wonderful Effects of the New Inoculation! — Vide. the Publications of S Anti Vaccine Society.

Edward Jenner's smallpox vaccination was initially ridiculed. This cartoon suggests people vaccinated with cowpox would grow cow-like features. (Library of Congress, Prints and Photographs Division, Cartoon Prints, British. LC-DIG-ds-14062)

Smallpox, meanwhile, has been eradicated (see entry 52). In 1959, the World Health Organization began a global smallpox eradication program. Eradication was believed to be possible because humans are the only species susceptible to the virus; there is no animal reservoir. The program got off to a slow start but was revitalized in 1967.

Smallpox was eradicated in the United States, and routine vaccinations were stopped in 1972. In 1975, the last person died of naturally acquired smallpox caused by *variola major*. The last case of naturally acquired smallpox caused by *variola minor* occurred in Somalia in 1977. The last person known to die from smallpox was Janet Parker, a British microbiologist, who died from laboratory-acquired smallpox in 1978. The world was officially declared smallpox-free on May 8, 1980.

Today a stockpile of smallpox virus exists only in freezers at the Centers for Disease Control and Prevention in Atlanta, Georgia, and at the State Research Center of Virology and Biotechnology in Koltsovo, Russia. These stockpiles remain controversial because of the potential for smallpox to be used as a bioterrorism weapon. The United States still makes a highly controlled supply of smallpox vaccine. Only a few people, such as smallpox researchers and some high-risk military members, are vaccinated each year.

FURTHER INFORMATION

Davidson, Tish. *Vaccines: History, Science, and Issues.* Santa Barbara, CA: Greenwood, 2017.

"History of Smallpox." Centers for Disease Control and Prevention, July 12, 2017. https://www.cdc.gov/smallpox/history/history.html

Riedel, Stefan. "Edward Jenner and the History of Smallpox and Vaccination." *Proceedings Baylor University Medical Center* 18, no. 1 (January 2006): 21–26. https://www.ncbi.nlm.nih.gov/pmc/articles/PMC1200696

12. First Successful Human-to-Human Blood Transfusion, 1829

The first successful human-to-human blood transfusion was performed by British obstetrician James Blundell (1790–1878) in 1829. Blundell was raised in London. His father owned a prosperous company that sold cloth and sewing supplies, but rather than join the family business, he followed his uncle into medicine. Blundell attended the University of Edinburgh in Scotland and received a medical degree in 1813. He then returned to London where he specialized in midwifery and obstetrics, joining his uncle as a co-lecturer in these subjects at St. Guy's Hospital School. Upon the death of his uncle in 1823, he became the school's only lecturer in pregnancy and childbirth.

Blundell became interested in blood transfusion as a way to treat postpartum hemorrhage after he saw many women bleed to death during childbirth. He began transfusion experiments in 1818 by trying to transfuse blood between dogs. First, he bled one dog until it was near death. Next, he opened an artery in a donor dog and collected this dog's blood. Using a syringe, he injected the donor dog's blood into a vein of the dying recipient dog. His early attempts led to the death of both the donor and recipient dogs. Gradually, Blundell refined his technique. He discovered that to prevent the donated blood from clotting, the transfusion had to be performed quickly and all air had to be eliminated from the syringe transferring the blood.

Once he perfected his technique enough that some dog-to-dog transfusions were successful, he tried transfusing human blood into dogs. All the dogs died. He concluded that despite earlier reports of animal blood being given to humans who then survived, large quantities of blood from one species could not successfully be given to another species.

The next step was for Blundell to try a human-to-human transfusion using the syringe method. He treated six patients, two of whom were hemorrhaging after child birth and one each with childbirth fever, a ruptured blood vessel, cancer, and hydrophobia (a symptom of rabies). None of the people improved. He reported these procedures in the medical journal *The Lancet* and continued to experiment with the procedure. His first successful human-to-human transfusion occurred in 1829 when he gave a woman bleeding after childbirth blood taken from her husband. The woman recovered.

Blundell recognized that air was the enemy of successful transfusions because it caused blood to clot. He developed a device called the Gravitator to facilitate direct transfusion between donor and recipient. The donor, standing above the recipient, had a tube inserted into an artery while the recipient had a tube inserted in a vein. The role of the Gravitator was to regulate the stream of blood being transferred and to prevent it from being exposed to air. The device was a success and was manufactured commercially although transfusion remained a relatively uncommon procedure.

Blundell went on to develop improved surgical techniques related to female reproductive anatomy and childbirth. His teaching lectures on childbirth procedures and the reproductive diseases of women were secretly transcribed in shorthand and published against his wishes. Blundell resigned from his teaching position in 1836 after a conflict with the university but continued to practice medicine privately until 1847. He lived another 31 years and died an extremely rich man.

HISTORICAL CONTEXT AND IMPACT

James Blundell was not the first person to experiment with blood transfusion. In 1666, British physician Richard Lower (1613–1691) published a description of transfusing blood between dogs in *Philosophical Transactions*, Europe's oldest scientific journal. Later, Lower worked with physician Edmund King (c. 1630–1709) to transfuse a man with blood from a sheep. The man survived. It was later thought that his survival was due to the tiny amount of blood transferred, as other animal-to-human transfusions consistently failed.

Around the same time in France, Jean-Baptiste Denys (1643–1704) transfused calf blood into Antoine Mauroy, a mentally ill man, who died after the third transfusion. Denys was charged with murdering Mauroy but was acquitted. He stopped practicing medicine, and human transfusions were outlawed in France. Although they remained legal in England, there was little enthusiasm for more human transfusion experimentation after Denys's experience with the law.

One reason Denys chose a mentally ill man as as a recipient for transfusion was to explore questions raised by Robert Boyle (1627–1691), a famous chemist, physicist, and philosopher. Shortly after Lower published a description of his dog-to-dog transfusion experiments, Boyle published a series of questions in the same journal about the effects of transfusion. He asked whether transfusion between different breeds of dog could change the recipient dog's breed. Could it change a dog's temperament? Would transfusion have an effect on learned behavior or cause the dog to not recognize its owner? Denys appeared to be aware of these questions and hoped to show that blood transfusion could cure his recipient of mental illness. Instead, it killed him.

A breakthrough in understanding transfusion came in 1901 when Austrian physician Karl Landsteiner (1868–1943) discovered that red blood cells have proteins, called antigens, on their surface. He detected three different types of blood antigens, which he called A, B, and C. Type C is now called type O. The following year, a fourth blood group, AB, was discovered by two of Landsteiner's students.

Transfusion of incompatible blood types can be fatal. The immune system attacks the incompatible cells causing clumping and cellular destruction. In terms of compatibility, AB blood can receive blood from any group, and O blood can be given to any blood group. Types A and B must receive an identical type or type O. Landsteiner received the Nobel Prize in Physiology or Medicine in 1930 for his discovery. Rh factor, the second most familiar blood antigen, was not identified until 1939. Rh is a protein that is either present (Rh positive, or Rh+) or absent (Rh negative, or Rh−) on blood cells. People with Rh negative blood cannot safely receive Rh positive blood. Additional blood antigens continue to be identified and are important in matching organ donors to recipients.

Blood does not keep well. For many years, if someone needed a transfusion, the doctor had to do a quick search for a person with the correct blood type who was willing to do a direct person-to-person transfusion using almost the same techniques that Blundell used. Gradually, methods to temporarily store blood were developed. In 1916, Francis Peyton Rous (1879–1970) and Joseph R. Turner found that by adding a citrate-glucose solution to blood, the blood could be refrigerated and remain good for indirect transfusions for several days. This allowed Oswald Robertson (1896–1966), an American army officer, to create the first "blood depots" during World War I. During peacetime, the concept of a blood depot was transformed into what we now recognize as a blood bank. The first blood bank was established in Leningrad in 1932. The first American hospital blood bank was created at the Cook County Hospital in Chicago in 1937.

With the ability to store blood came a question similar to that asked by Boyle in the 1600s. Would transfusion of blood from a Black person into a white person make that person Black? This was of particular importance in the Deep South where the one-drop rule meant that a single drop of "Black blood" made a person a Black. The one-drop rule resulted in segregation of blood and blood products by race, a procedure not banned in Louisiana until 1972.

World War II increased the need to make blood available on the battlefield. Blood contains multiple components—red and white cells, platelets, and plasma. Plasma consists of water, salt, enzymes, and proteins. Charles Drew (1904–1950), an African American researcher who received his medical education in Canada, discovered that plasma could be separated from the cellular part of blood and dried into a powder that could be kept for a long time. The powder could be reconstituted using a special fluid and often could save lives even without the

James Blundell performed the first successful human blood transfusion by connecting the donor directly to the recipient. (Wellcome Collection. Attribution 4.0 International (CC BY 4.0))

cellular component. In 1940, Drew was put in charge of shipping dried plasma to Britain for use on the battlefield. The following year he became the first director of the American Red Cross and was responsible for supplying blood and plasma to the American army and navy. However, the government decreed that Black blood should be kept separate from white blood so that combatants could receive blood only from their own race. This angered Drew so much that he resigned from the Red Cross. The Red Cross continued to segregate blood by race until 1950, and the practice continued much longer in some hospitals.

Today transfusions are common, although with an increase in laparoscopic surgery (see entry 29), less blood is used because laparoscopic surgery causes less bleeding According to the American Red Cross, about 36,000 units of whole blood, 10,000 units of plasma, and 7,000 units of platelets were transfused in the United States every day in 2019.

FURTHER INFORMATION

AABB. "Highlights of Transfusion Medicine History." Undated. https://www.aabb .org/news-resources/resources/transfusion-medicine/highlights-of -transfusion-medicine-history

Hedley-Whyte, John, and Debra T. Milamed. "Blood and War." *Ulster Medical Journal* 79, no. 3 (September 2010): 125–134. https://www.ncbi.nlm.nih.gov/pmc/articles /PMC3284718

National Museum of African American History. "The Color of Blood." Undated. https:// nmaahc.si.edu/blog-post/color-blood

Yale, Elizabeth. "First Blood Transfusion: A History." *JSTOR Daily*, April 22, 2015. https:// daily.jstor.org/first-blood-transfusion

13. First University-Trained African American Doctor, 1837

James McCune Smith (1813–1865) was the first African American to earn a full medical degree from a university. Smith was born in New York City on April 18, 1813. His mother was what Smith in his adult life called a "self-emancipated Black woman." His father, a white man, was never part of his life.

As a child, Smith excelled as a student at African Free School #2. African Free Schools were established—beginning in 1794—by the New York Manumission Society to educate children of enslaved persons and people of color. They operated on the premise that Black children were as intelligent and capable of learning as white children. Through this school, Smith acquired an excellent education as well as connections to many prominent abolitionists.

Upon graduation, Smith applied to study medicine at both Columbia University and Geneva Medical College in Geneva, New York. Despite an outstanding scholastic record, he was rejected by both schools because of his race. Through wealthy abolitionists connected to the African Free School and their relationship with the Glasgow Emancipation Society, Smith was admitted to Glasgow University in Scotland. Here, between 1832 and 1837, he earned a Bachelor of Medicine degree, a Master of Medicine degree, and a Doctor of Medicine degree. After graduation, he furthered his education by going to Paris for additional clinical work.

Upon returning to New York City in 1837 as a fully qualified, university-trained African American physician, Smith opened a medical and surgical practice. He also started an apothecary shop that sold herbs. He is thought to be the first Black man to own such a store. Smith also served as the physician for the Colored Orphan Asylum for 20 years until the orphanage was burned down by an angry white mob during the New York City draft riots in July 1863.

Smith applied for membership in the New York Medical and Surgical Society but, despite finding him qualified, the Society denied him membership because of his race. Later, he would become the first Black doctor to write a medical case report. Because he was not a member, he could not read the report before the

Society. Instead, his report was read by a white colleague who had consulted on the case. A second case report written by Smith was published in the *New York Journal of Medicine*. It is believed to be the first scientific medical paper published in America by a Black man.

Smith's activities were not limited to medicine. A skilled statistician, he used his abilities to refute ideas about the health and well-being of Blacks that arose from the racially biased census of 1840. In 1854, he was elected to the American Geographical Society in New York where he used his knowledge of statistics to outline ways to improve the methods used in conducting the census.

Smith also participated actively in abolitionist activities on the national level, often writing eloquently about discriminatory perceptions and policies. He served as the director of the Colored People's Educational Movement and collaborated with famous abolitionists such as Frederick Douglass (1818–1895) and Gerrit Smith (1797–1874). He helped Douglass found the National Council of Colored People and was active in founding the Legal Rights Association of New York, which focused on minority rights and served as a model for the National Association for the Advancement of Colored People.

Smith was financially successful. He married Malvina Barnet, reportedly a free woman of color and a college graduate. They had 11 children of which five survived to adulthood. Smith had a mansion built for his family. In 1860, it was worth about $25,000 ($800,000 in 2022 dollars). The family lived in several different neighborhoods in New York City. Interestingly, when they lived in a Black neighborhood, they were listed on census documents as mulatto. When they lived in a white neighborhood, they were listed as white. All four of Smith's surviving sons married white women and passed into white society.

HISTORICAL CONTEXT AND IMPACT

James McCune Smith was the first African American with a university medical degree to practice medicine in the United States, but there was another African American who practiced medicine in the early 1780s. He was apprentice-trained as was common at the time. James Derham, sometimes spelled Durham (1762–?), was born enslaved in Philadelphia. At an early age, he became the property of Dr. James Kearsley and learned about medicine by assisting him. During the Revolutionary War, Kearsley was arrested for treason and died in prison. Derham then became the property of George West, a surgeon, and later Dr. Robert Dow, from whom he bought his freedom.

Derham became friends with Benjamin Rush, signer of the Declaration of Independence and one of the most famous physicians of the time. In 1788, Derham opened a financially successful practice in New Orleans, earning $3,000 a year (about $100,000 in 2022). While in New Orleans, he continued to correspond with Rush. His last known letter is dated April 5, 1802, after which he disappears from the historical record.

David Jones Peck (c. 1826–1855), born in Carlisle, Pennsylvania, was the first African American to graduate from an American medical school. He graduated from Rush Medical College in Chicago, Illinois, in 1847 and established an unsuccessful medical practice in Philadelphia. In 1852, he and his wife moved to Nicaragua as part of a group of Black immigrants who established the town of San Juan Del Norte. Peck was chosen mayor and soon became involved in the Nicaraguan Civil War. In 1855, he was killed by cannon fire in the city of Grenada where he is buried.

History shows that getting professional recognition as a Black doctor has always been difficult. Professional societies such as the American Medical Association (AMA) denied membership to Black doctors regardless of how qualified they were. For years, the AMA left it up to state organizations to determine whether Black members should be admitted. Thus, John V. DeGrasse became the first African American to be admitted to the Massachusetts Medical Society in 1854 while Southern state medical societies were denying membership to Black doctors well into the 1960s. Some African American doctors formed local or state-wide Black medical organizations, but the AMA unapologetically refused to recognize delegates from these societies at its national meetings.

Membership in professional medical societies is linked to professional advancement, hospital admitting privileges, patient referrals, and training in the latest medical advances. Members have a forum to present papers and can call on colleagues for advice and assistance. Black doctors who were excluded from medical societies had none of these advantages. They remained isolated and less well informed on current practices.

As a result of their exclusion from the AMA and its state and local branches, African American doctors founded the National Medical Association (NMA) in 1895 to support Black health professionals. Notably, the organization had no race-based membership criteria. The NMA's focus was on eliminating health-care disparities faced by minority populations, improving minority health outcomes, and providing its members with the advantages of professional group membership. The NMA struggled to survive during its early years and was generally ignored by the AMA. It successfully published its first professional medical journal in 1909.

In addition to working to improve medical training for Black doctors and medical care for disadvantaged patients, in the years following World War II, the NMA was active in increasing the number of Black medical students. Between 1948 and 1956, the percentage of African Americans enrolled in medical schools increased significantly. The majority of Black medical school graduates came from Howard University College of Medicine and Meharry Medical College, both well-respected historically Black institutions. Yet, despite some progress, by 1960 only 14 out of 26 medical schools in the South admitted Black students. This inspired the NMA to take an active role in lobbying for passage of the Civil Rights Act and in voter registration drives in the South.

In 2008, the AMA formally apologized to the NMA and Black physicians for its discriminatory policies. However, the results of this discrimination remain. In 2018, African Americans made up 13% of the U.S. population but just over 4% of all practicing physicians. The number of Black female doctors has increased faster than Black male doctors. In 2015, medical schools were graduating 31% more Black female doctors than Black male doctors.

FURTHER INFORMATION

Harris, Michael J. "David Jones Peck, MD: A Dream Denied." *Journal of the National Medical Association* 88, no. 9 (September 1996): 600–604. https://www.ncbi.nlm.nih.gov /pmc/articles/PMC2608104

Institute for Ethics at the AMA. "The History of African Americans and Organized Medicine." *Journal of the American Medical Association* 300, no. 3 (2008): 306–313. https://www .ama-assn.org/about/ama-history/history-african-americans-and-organized-medicine

Morgan, Thomas M. "The Education and Medical Practice of Dr. James McCune Smith (1813–1865), First Black American to Hold a Medical Degree." *Journal of the National Medical Association* 95, no. 7 (July 2003): 603–614.

Wynes, Charles E. "Dr. James Durham, Mysterious Eighteenth-Century Black Physician: Man or Myth." *Pennsylvania Magazine of History and Biography* 103, no. 3 (July 1979): 325–333.

14. First Hypodermic Needle, 1844

The first hypodermic needle was created in 1844 by Irish physician Francis Rynd (1801–1861). Rynd was born in Dublin to a wealthy family. He received his medical education at Trinity College, Dublin. He was an indifferent student who preferred foxhunting and parties to studying. Nevertheless, after graduating, he was apprenticed to Sir Philip Crampton (1777–1858), a prominent surgeon at Meath Hospital, Dublin, who served as president of the Royal College of Surgeons in Ireland. Rynd married the daughter of the Lord Mayor of Dublin and maintained an active social life in Dublin's high society.

Despite an undistinguished academic career and busy social life, Rynd became a dedicated physician and, in 1830, a member of the Royal College of Surgeons. He had a large private practice, was the medical superintendent of Mountjoy Prison, and worked as a surgeon at Meath Hospital until his death from a heart attack.

Although today hypodermic needles have many other uses, the original purpose of Rynd's hypodermic needle was to place drugs under the skin (*hypo* meaning "below" and *derma* meaning "skin"). Rynd's idea was that drugs, usually opium-based painkillers intended as local anesthetics, would act faster and more effectively if injected near the site of the pain rather than taken by mouth or applied to the skin.

Rynd's hypodermic needle was not a needle as we think of it today. It consisted of rolled metal drawn into a very thin hollow tube that was open at both ends. Inserted into one end of the cannula was a triangular piece of metal with three sharp edges, called a "trocar," a word derived from the French for three (*trois*) and sides (*carre*).

First, the skin was pierced with a lancet and the cannula inserted into the incision. Pressing a lever on the side of the apparatus caused the trocar to emerge from the cannula, rather like clicking a ballpoint pen causes the writing point to appear. A drug solution was inserted into a hole at the base of the cannula. Gravity then pulled this liquid into the body. Rynd did not make his own apparatus. It was made to his specifications by John Weiss, a skilled surgical-instrument maker. As of 2023, John Weiss & Sons still exists and continues to make surgical instruments.

Rynd first successfully used his hypodermic needle on June 3, 1844, according to an account he published in the March 12, 1845 issue of the *Dublin Medical Press*. He treated a 59-year-old woman who had extreme neuralgia (nerve pain) on the left side of her face by giving her morphine dissolved in creosote that he introduced under the skin near the facial nerve using the hypodermic needle. Almost immediately, the pain disappeared. The pain recurred in a milder form one week later, so the procedure was repeated. After that, the pain permanently disappeared. What Rynd had done was introduce a local anesthetic similar to the way a dentist injects an anesthetic before pulling a tooth.

HISTORICAL CONTEXT AND IMPACT

From ancient Greek and Roman times, various unsuccessful attempts were made to inject substances into animals and humans. In Europe, Christopher Wren (1632–1723) experimented with ways to inject fluids into dogs before he became a famous architect. The dogs all died, suggesting architecture was a more promising career choice for Wren. Others in the mid-1600s tried experiments on both animals and humans with similar dismal results. For the most part, injection experiments were then abandoned for 200 years until technology advanced to the point where an effective device could be manufactured.

Although Rynd was successful in developing the first hypodermic needle, a major limitation of his apparatus was that it depended on gravity to pull the drug solution into the body. Blood flowing through the circulatory system creates pressure in arteries and to a lesser degree in veins. Rynd's gravity-dependent hypodermic needle could not overcome this pressure, so it could not inject drugs into the circulatory system where they would be carried to distant parts of the body. His hypodermic needle was limited to injections just under the skin, a problem that was solved in the same year by two physicians who were unknown to each other.

Alexander Wood (1817–1884) was a Scottish physician who received his medical education at Edinburgh University and then established a local medical practice. Wood was interested in anesthesia, and in 1853, he developed a syringe that would push fluid through a hypodermic needle. This allowed anesthesia to be delivered more effectively and, later, to be delivered into a vein whose resistance would have prevented injection by a gravity flow device.

Wood's syringe was made of glass to which a hypodermic needle was attached. It used a plunger wrapped in waxed linen to deliver the fluid. The plunger was operated by a screw rather than the push-pull method used today. The glass syringe allowed the physician to see how much fluid was injected, which resulted in more accurate dosing. Wood tried to claim that he invented the hypodermic needle, but Rynd's report in the *Dublin Medical Press* took precedence. Wood was, however, one of the first to use a syringe to deliver drugs through the circulatory system. This greatly expanded the usefulness of the hypodermic needle.

Also in 1853, unknown to Wood, the French surgeon Charles-Gabriel Pravaz (1791–1853) developed a syringe made of silver to deliver fluids through a hypodermic needle. He was the first to record injecting a drug into a damaged artery in an attempt to make blood clot. Pravaz died soon after his syringe experiments, and the glass syringe developed by Wood became the widely used version. Before the germ theory became accepted and the need for sterility understood, syringes and needles were reused and could spread diseases.

The next big advance came in the early 1900s when H. Wulfing Luer, an instrument maker in Paris, replaced the screw mechanism controlling the plunger with a push-pull plunger on a glass syringe. These Luer syringes were first manufactured in the United States by Becton, Dickinson and Company in 1906. This company, now known as BG, still makes medical equipment for worldwide distribution.

Luer also made significant changes to the way the needle was attached to the syringe that are reflected in the design of syringes used today. Other changes were gradually incorporated into syringe design, but the needle remained essentially the same until 1922 when the mass production of insulin, a lifesaving treatment for diabetes (see entry 33), became available. Insulin must be injected under the skin; it is ineffective if taken by mouth. Thousands of diabetics injecting themselves with insulin multiple times daily spurred the development of a variety of hypodermic needles designed for specialized uses.

Changes in syringes were slower to develop. Blood shortages during the Korean War (1950–1953) stimulated the development of sterile syringes for collecting blood, along with additional specialized needles large enough to prevent damage to blood cells. In 1949 and 1950, an inventor named Arthur E. Smith received eight patents on a partially disposable glass syringe. The syringe went into mass production in 1954 and was used for the massive polio immunization campaign that occurred after the Jonas Salk polio vaccine was approved in April 1955.

Glass syringes were supposed to withstand about 150 hours of heat sterilization over their lifetime, but sterilization required equipment and took time. Concern about cross-contamination by reuse of syringes and needles in the 1950s led to the first fully disposable plastic syringe. It was developed in 1956 by Colin Murdoch (1929–2008), a New Zealand veterinarian and pharmacist. This syringe was not widely used until the early 1960s, partly because of its cost. It remains the basis for syringes used today. The AIDS epidemic of the 1980s reinforced concern about cross-contamination by reused needles and led to a number of new procedures to protect healthcare workers handling needles, or sharps as they are called today.

In 2021, fully disposable plastic syringes with metal needles became standard in all developed countries. Glass syringes are used in rare cases where the drug administered could interact with plastic. Hypodermic needles have become highly specialized, with different specifications for needles used in administering drugs and in collecting blood samples. The diameter of the needle, called the gauge, ranges from 7, the needle with the largest diameter, to 33, the needle with the smallest diameter. Needle length also varies with its specific use.

Needle phobia, sometimes called trypanophobia, is estimated to affect between 3.5% and 10% of people. It is most common in children over age five. Some people faint during medical procedures involving injections. Researchers think this response, called a vasovagal reflex reaction, is physiological and probably genetic in origin rather than a psychological or mental health issue.

FURTHER INFORMATION

Craig, Robert. "A History of Syringes and Needles." University of Queensland Faculty of Medicine, December 29, 2018. https://medicine.uq.edu.au/blog/2018/12/history-syringes-and-needles

Emanuelson, Jerry. "The Needle Phobia Page." 2020. https://futurescience.com/needles

"How It's Made Hypodermic Needles." YouTube, January 14, 2022. https://www.youtube.com/watch?v=gh9kgA0cENA

Mogey, G. A. "Centenary of Hypodermic Injection." *British Medical Journal* 2, no. 4847 (November 28, 1953): 1180–1185. https://www.ncbi.nlm.nih.gov/pmc/articles/PMC2030174

15. First Use of Anesthesia in Surgery, 1846

Anesthesia is artificially induced insensitivity to pain. The word was coined by Boston physician and poet Oliver Wendell Holmes, Sr. (1809–1894) from two Greek words meaning "without sensation." Before the use of anesthesia, surgery was an agonizing procedure to be avoided whenever possible. The first use of anesthesia in surgery, however, remains in dispute.

In 1800, British chemist Humphry Davy (1778–1829) discovered that inhaling nitrous oxide caused euphoria, calmness, and laughter. He called the gas "laughing gas." Although Davy experimented on himself with the gas—as a hangover cure—he did not pursue its use as an anesthetic.

By the late 1830s in America, it was fashionable to attend traveling lectures called "ether frolics." At these events, people in the audience were given nitrous oxide gas or diethyl ether in order to experience their mind-altering and pain-inhibiting effects. All the men who claimed to be the first to use these gases as anesthetics were introduced to them at ether frolics.

Dr. Horace Wells (1815–1848) discovered the pain-suppressing properties of nitrous oxide when he attended a frolic and observed a man who had injured himself while under its influence but who did not react to the pain. Wells was a dentist. At age 19 he began a two-year apprenticeship in dentistry, the common way to become a dentist at that time. Following his apprenticeship, Wells set up a practice in Connecticut, where he first encountered nitrous oxide. After seeing its effects at a frolic, he persuaded another dentist to extract one of his teeth while he was under the influence of the gas. When the gas wore off, he reported that he had felt no pain.

Wells proceeded to use the gas on his patients, and in 1845, he persuaded Harvard Medical School to let him give a demonstration of painless tooth extraction. Unfortunately, the patient screamed throughout the procedure. At that time, it was not known that involuntary screaming could be a side effect of the gas. The next day the patient claimed to have felt nothing, but Wells was still declared a fraud.

Wells appears to have been an unstable person, opening, closing, and relocating his dental practice multiple times. Eventually, he abandoned his family and went to New York City where he became addicted to sniffing ether and chloroform. In January 1848, apparently under the influence of drugs, he threw sulfuric acid on two prostitutes. He was arrested and jailed but persuaded the jailer to let him return home under guard to collect some personal items. While there, he used his razor to open an artery and commit suicide.

A few days before his death, the Parisian Medical Society had honored Wells as the first person to perform a surgical operation without pain. In 1864 the American Dental Society recognized Wells as the discoverer of modern anesthesia, and the American Medical Association recognized this achievement in 1870.

William T. G. Morton (1819–1868) also claimed to be the first person to use anesthesia in surgery. Morton graduated from the Baltimore College of Dental Surgery in 1840 and then went to study with Horace Wells in Connecticut. In 1844, Morton entered Harvard Medical School. While at Harvard, he performed a painless tooth extraction. Following this, Morton persuaded surgeon John Collins Warren (1778–1856) to allow him to anesthetize a patient, Edward Abbott, so that Warren could remove a tumor from Abbott's neck. The surgery was publicly performed on October 16, 1846, at Massachusetts General Hospital.

Word of the painless operation spread quickly. Morton published the results of the surgery but tried to conceal the gas he had used by naming it letheon and claiming that it was a gas of his own invention rather than commonly available ether. He also claimed to be the first person to anesthetize a patient. These claims were promptly disputed by Harvard chemistry professor Charles T. Jackson and dentist Horace Wells.

Until well into the twentieth century, the medical profession considered it unethical to profit from medical discoveries. Morton, however, was determined to make money from the procedure. He first tried unsuccessfully to get a patent on the gas he called letheon. Next, he alienated his professional colleagues by applying to the Congress for $100,000 for his "discovery," but the claim went nowhere because of conflicting claims by Wells and Jackson. Morton reapplied to Congress in 1849, 1851, and 1853 and then tried to sue the federal government. All attempts to gain financial compensation failed; however, in 1871, a group of private citizens published a national testimonial proclaiming Morton the inventor of anesthetic inhalation. In 1944, Paramount pictures made a movie about Morton's life called *The Great Moment*, enshrining Morton in the minds of the public as the first person to use anesthesia in surgery.

Crawford Long (1815–1878), a Georgia surgeon, is the earliest candidate for the first use of anesthesia in surgery but the one with the least documentation. Crawford graduated from the University of Georgia and then received a medical degree from the University of Pennsylvania. He chose to return to Georgia to practice medicine in the small town of Jefferson. There, on March 30, 1842, he removed a cyst from the neck of James Venable using ether as an anesthetic. He later performed eight more surgeries, including the amputation of a finger, before Morton's public demonstration of anesthesia in 1846.

Long, however, did not publish his results until 1849, by which time Morton had become in the eyes of the public the first person to use anesthesia during surgery. Long was also the first person to use anesthesia during childbirth, having anesthetized his wife during the birth of their second child. This accomplishment was not publicized, and Dr. James Simpson of Edinburgh, Scotland, claimed this honor.

The best explanation of why Long did not publish his findings is that he wanted to perform multiple surgeries to solidify his results, but since he practiced medicine in a rural county, he only had a few surgical cases each year. Nevertheless, in 1879, the National Eclectic Medical Association, an organization that promoted botanical remedies and Native American medical practices, named Long the official discoverer of anesthesia. In 1940, a U.S. postage stamp was issued in his honor, and a plaque at the University of Georgia recognizes his achievement.

HISTORICAL CONTEXT AND IMPACT

For centuries, cultures tried to find a way to dull the pain of surgery. As early as 1600 BCE the Chinese used acupuncture, but most remedies involved a mixture of herbs and alcohol. Opium, known since ancient times, and morphine, isolated in 1805 by Friedrich Sertürner (1783–1841), were commonly used for pain control, but nothing was as effective as anesthetic gas.

Using anesthetic gas to render the patient unconscious eliminated the horrific pain of surgery and immobilized the body so that it was no longer necessary to tie the patient down for the surgeon to operate. This permitted slower, more accurate surgery with better results. As the news of pain-free surgery spread, more people were willing to undergo potentially lifesaving operations that had earlier been avoided.

In 1881, the world's first medical journal devoted solely to anesthesia was published. Two years later, the first professional anesthesia society was formed in London. The first American anesthesia society, the Long Island Society of Anesthetists, was not established until 1905. It became the American Society of Anesthetists in 1936. Anesthesia became a board-certified specialty in 1940. As of 2021, according to the U.S. Bureau of Labor Statistics, about 31,300 anesthesiologists were practicing in the United States with an average annual wage of $330,000.

The first anesthesia was gas anesthesia, and there was a fine line between anesthetizing the patient and death by overdose, but in 1884 Austrian Karl Koller (1857–1944) introduced cocaine as an anesthetic for local eye surgery. This ushered in the beginning of a safer era for anesthetics. Today there are three common kinds of anesthesia: general anesthesia, where the entire body is numbed; regional anesthesia to block pain in a specific region of the body; and local anesthesia, where only a small part of the body is numbed. In addition, there is conscious sedation, sometimes called twilight sedation, in which the patient is sedated and relaxed but remains conscious. Conscious sedation is commonly used for cataract surgery along with a topical anesthetic that numbs the eye. General anesthesia and conscious sedation can be administered by intravenously drip. Regional anesthesia is administered by injection once the patient is lightly sedated. Local anesthesia, such as that used to extract a tooth, is given by injection.

Researchers today understand that local and regional anesthetics block nerves to prevent them from carrying pain signals to the brain. It is still unclear how modern general anesthetics work. One 2019 research study found that in mice, rather than block nerve transmission, general anesthetics activate changes in a group of cells located deep in the brain that regulate mood, sleep, and basic body functions. This area of research continues to be explored.

Modern anesthesia is quite safe. In the 1970s, 357 people per million died from anesthesia. By 2000, the number had dropped to 34 per million, with more than 230 million major surgeries occurring around the world each year. As of 2021, the increased use of laparoscopic surgery, which requires a much smaller incision, has substantially decreased the use of general anesthesia.

FURTHER INFORMATION

Aptowicz, Cristin O'Keefe. "The Dawn of Modern Anesthesia." *The Atlantic*, September 4, 2014. https://www.theatlantic.com/health/archive/2014/09/dr-mutters-marvels/378688

Thomas, Roger K. "How Ether Went from a Recreational 'Frolic' Drug to the First Surgery Anesthetic." *Smithsonian Magazine*, March 28, 2019. https://www.smithsonianmag .com/science-nature/ether-went-from-recreational-frolic-drug-first-surgery-anesthetic -180971820

Wood Library-Museum of Anesthesiology. "History of Anesthesiology." 2020. https:// www.woodlibrarymuseum.org/history-of-anesthesia

16. First International Health Organization, 1851

The International Sanitary Conferences (ISC) was the first formal organization to tackle problems of international health. It was established in 1851 by the French minister of agriculture and trade in response to a cholera pandemic sweeping across Europe. The initial goal of the ISC was to establish uniform minimum maritime requirements for quarantining ships carrying people sick with cholera. Quarantine regulations for yellow fever and plague were also on the agenda, but they received little attention compared to the effort put into stopping the spread of cholera.

Cholera originated in the Ganges Delta in India, where regional outbreaks occurred frequently. The first recorded cholera pandemic began in 1817 and lasted until 1824. The disease spread rapidly along trade routes from India to Southeast Asia, the Middle East, eastern Africa, and the Mediterranean coast. Hundreds of thousands of people died. A second pandemic began around 1829. Cholera reached Russia and virtually all of Europe by the early 1830s. After that, periodic epidemics occurred driven by increased rail and ship trade and a growing population crowded into cities where sanitation was poor or nonexistent.

The first meeting of the ISC began on July 23, 1851. Eleven European governments and Turkey sent two representatives each, one a physician, the other a diplomat. Each representative got one vote and was not required to vote the same way as his fellow countryman, an arrangement that often led to gridlock. Representatives were also handicapped because the way cholera was transmitted was unknown at the time of the first meeting (see entry 17).

Cholera, we now know, is an intestinal infection caused by the bacterium *Vibrio cholerae*. It is spread by consuming food or water contaminated by feces from infected individuals. It is not directly transmitted from person to person. The disease causes vomiting and severe watery diarrhea. An infected person can lose up to one quart (about 1L) of fluid per hour or about five gallons (about 20L) per day. Infected people quickly die from dehydration. Localized outbreaks of cholera still occur, especially after natural disasters such as the 2010 earthquake in Haiti.

Even in modern times, about one out of every 20 people infected will die. The rate was much higher in the 1800s.

Delegates to the 1851 ISC meeting had varying opinions about cholera and how to respond to it. One group claimed that cholera was caused by environmental factors ("bad air"). A second group believed that cholera, yellow fever, plague, malaria, and typhus were all the same disease but that the disease took different forms in different people. A third group insisted that cholera was not transmitted from person to person. All three groups opposed the idea of quarantine as a useless and needless intervention. The Austrian delegate even claimed that his country had tried quarantine and that it made the disease worse.

Other delegates wanted strong quarantine regulations. They felt quarantines had helped slow the spread of other diseases and would help slow the spread of cholera. Despite presenting various health positions, the underlying issue was how quarantine rules would affect trade and the economy. Countries such as England whose economy was heavily dependent on international sea trade wanted no quarantine regulations because quarantine caused loss of time and money. In response, the Spanish medical delegate stated that those who opposed quarantine regulations were elevating money and profits over public health. In the end, a resolution was passed making quarantine optional; in other words, nothing changed.

Despite the lack of action in their first meeting, the ISC continued to meet at irregular intervals. In all, 14 meetings occurred over a span of 87 years. In time, additional countries sent delegates. Many of the meetings lasted for months. Most were held in Paris. Only the fifth meeting, held in 1881, occurred in the United States.

The first international sanitary agreements on cholera were finally passed at the 1892 meeting in Venice, 41 years after they were first proposed. At the eleventh meeting in 1903, the quarantine agreement was extended to include yellow fever and plague. The 1926 meeting expanded what was known as the International Maritime Sanitary Convention. The last ISC meeting was held in Paris on October 28, 1938, with almost 50 countries participating.

HISTORICAL CONTEXT AND IMPACT

Before the first attempts to develop an international approach to disease prevention, pandemics were handled at the local or regional level. Many countries in Europe were not fully unified. For example, at the time of the first ISC meeting, delegates were sent from the Papal States, Sardinia, Tuscany, and the Kingdom of the Two Sicilies, all of which later became part of Italy.

The one institution that reached across countries was the Church. The Church, both Protestant and Catholic, connected illness with sin and saw it as a divine judgment from God. In the Church's view, if you got sick it was because you had displeased God. The recommended cures were prayer for divine intervention and fasting. The Church sometimes saw treatment advised by physicians as a threat to its power.

Another factor that kept pandemic responses at the local level was that communication was slow and limited while disease traveled rapidly, so rapidly that regions had little chance to learn from each other's experiences. Nevertheless, as international trade and travel increased, there began a movement to develop regional health standards and regulations concerning specific diseases. This first resulted in the formation of the ISC and later in other international health organizations.

In 1902, the Pan American Health Organization was formed to deal with the health concerns of the Americas, especially yellow fever (see entry 27). The Office International d'Hygiène Publique (OIHP or International Office of Public Health), headquartered in Paris, grew out of a meeting of 11 countries in Rome in 1907. It expanded the idea of international health standards beyond quarantine measures to areas such as safe drinking water and sanitation.

The League of Nations, founded in 1920 after World War I (1914–1918) and headquartered in Geneva, Switzerland, was primarily interested in maintaining world peace. It did, however, establish the Health Committee that absorbed the OIHP into the Committee's somewhat complex and unwieldy structure. Later, as air transportation improved, international air travel became more common. A multinational meeting at The Hague in 1933 resulted in the passage of the International Sanitary Convention for Aerial Navigation.

The League of Nations proved ineffective in maintaining world peace. During World War II (1939–1945), most international health initiatives and agreements were disrupted or disregarded. When the war ended, a committee met in San Francisco to establish a replacement organization for the League of Nations. The result was the United Nations which came into existence on October 24, 1945 and represented 51 countries. As of 2022, the United Nations had 193 member states.

The World Health Organization (WHO) is a specialized agency of the United Nations. It was established on April 7, 1948 at the urging of delegates from China, Norway, and Brazil. Its stated objective is "the attainment by all people of the highest possible level of health."

The WHO absorbed all the previously recognized multinational health organizations associated with the League of Nations and took over their responsibilities. Over time, the organization added six regional offices serving the Americas, Europe, the Western Pacific, Africa, the Eastern Mediterranean, and Southeast Asia.

The WHO acts as both an advisor to governments and national health organizations and also sponsors specific health intervention programs. In its advisory role, it has created a list of essential medicines. Countries can use this approved list to develop customized national drug lists to best maintain the health of their citizens. The drugs on the list are chosen based on safety, effectiveness, relative cost, and prevalence of the disease that the drug treats. Every two years the list is revised. In 2022, it contained 479 medications. The WHO-approved essential drug list is especially useful to countries that do not have robust independent drug-approval programs.

The WHO also sponsors specific intervention activities and partners with non-governmental organizations that support these activities. For example, it maintains an active worldwide immunization program to help control communicable diseases that are a universal threat. Its most successful immunization program eradicated smallpox in 1980 (see entry 52). A similar program has achieved the near eradication of polio, although an outbreak of polio in the New York City area in 2022 was a setback for the program. Outbreaks of cholera and yellow fever, two diseases that were the concern of the first health advisory organization, still occur, but less frequently and in more limited areas. Cholera has been reduced through education and sanitation projects in underdeveloped countries, and yellow fever has been controlled through mosquito abatement programs and vaccination.

In an effort to stay ahead of potential pandemics, the WHO maintains a research program on emerging diseases that could cause or are actively causing outbreaks. These include Zika, Marburg virus disease, Lassa fever, Rift Valley fever, and Middle East respiratory syndrome. Nevertheless, unexpected health challenges, such as the COVID-19 pandemic that began in 2020 and monkeypox outbreaks in 2022, still occur. During epidemics, the WHO provides up-to-date science-based advice to governments and citizens on how best to control the spread of the disease. It does not, however, have any legal authority to force countries to follow this advice. Just like the government representatives of the 1850s, today's government officials facing a pandemic must balance the damage to trade and the economy against vigorous support of public health programs.

FURTHER INFORMATION

Clift, Charles. "The Role of the World Health Organization in the International System." Chatham House Working Group on Governance Paper 1, February 2013. https://www.chathamhouse.org/sites/default/files/publications/research/2013-02-01-role-world-health-organization-international-system-clift.pdf

Howard-Jones, Norman. "The Scientific Background of the International Sanitary Conferences 1851–1936." World Health Organization, 1995. https://apps.who.int/iris/handle/10665/62873

Taylor, Allyn L. "International Law and the Globalization of Infectious Diseases." YouTube, May 10, 2020. https://www.youtube.com/watch?v=Q723v7RafJk

Your Daily Dose of Knowledge. "The History and Development of World Health Organization." YouTube, April 17, 2020.

17. First Epidemiological Study, 1854

Epidemiology is a branch of medicine that studies the frequency and distribution of health events in specific populations. Epidemiologists analyze who becomes sick with a similar set of symptoms and where and when the illness occurs. From this information, they try to determine the cause of the illness and how it is spread in order to prevent or control the disease.

John Snow (1813–1858), a British doctor, is considered the father of epidemiology for the way he investigated the spread of cholera in London in 1854 and applied statistical analyses to prove his findings. Snow was born in the York, England, to poor family. Because he showed exceptional ability in mathematics, his parents encouraged him to get an education. At the age of 14, he was apprenticed to a surgeon-apothecary. His first experience with cholera occurred four years later when he treated miners who were ill with the disease.

Snow moved to London to continue his education. In 1844, he earned a medical degree from the University of London. By 1849, he had joined the Royal College of Physicians of London. Snow maintained a lifelong interest in cholera, but he was also an accomplished surgeon. His ability in mathematics allowed him to develop a formula to accurately calculate the appropriate dose of anesthesia for his patients. He was held in such high esteem that he was chosen to anesthetize Queen Victoria during the birth of her eighth child.

Cholera, which is caused by the bacterium *Vibrio cholerae*, originated in the Ganges Delta in India. The first Indian cholera epidemic was recorded in 1817. The disease spread rapidly, and by 1831 the first outbreak was recorded in England. After that, periodic cholera epidemics occurred, especially in cities such as London where living conditions were crowded and sanitation poor.

Cholera, we now know, is transmitted by consuming contaminated water or food. Within a few days, it attacks cells in the intestine and causes severe watery diarrhea. An infected person can lose up to one quart (about 1L) of water an hour or five gallons (about 20L) each day. Along with water, the individual loses salts and minerals the body needs to function. Left untreated, an infected individual almost always dies quickly from dehydration.

A cholera outbreak occurred in London in 1848 that inspired Snow and other London physicians to establish the London Epidemiological Society. Its goal was to advise the government on how to combat the disease, but the outbreak ended before the Society had much impact.

Snow's real breakthrough came during the London cholera epidemic that exploded in 1854. In the 1850s, many physicians and the public believed that disease came from "bad air" or unpleasant odors. Snow rejected this idea, believing that living organisms, passed in some way from person to person, caused disease. This belief led him to study the spread of cholera in what became the first epidemiological study.

In 1854, London was a crowded city with a poor sanitation system. Human waste was collected in sewers and open ditches and dumped into the Thames River. Two companies, Southwark and Vauxhall (S&V) and Lambeth, supplied water. Snow discovered that Lambeth got its water miles above the point where sewage entered the river, while S&V collected its water from just below the sewage discharge sites. Both companies served the same population, often providing water to different houses on the same street.

When cholera broke out, Snow began his investigation by going door to door asking residents which company provided their water and how many people in their household had been infected with cholera. He was able to show statistically that people who got their water from S&V were 14 times more likely to get cholera than those who got their water from Lambeth. This strongly suggested that cholera was being transmitted by something in sewage-filled water. Snow then began another investigation in an area where people got their water from public wells. He made the assumption that people would get their water from the pump closest to their home. He then marked the location of the pumps and the houses with cholera deaths on a map. This revealed a startling correlation. Of 73 people who had died from cholera in one area, 61 were known to have drunk water from a single pump known as the Broad Street pump. Snow's statistics were strong enough to convince public officials to disable the pump. Later, it was found that water from a cistern contaminated with cholera-laden fecal waste had leaked into the Broad Street well.

Snow's first insight was that cholera was caused by an infectious organism and not by bad air. In 1854, around the same time that he determined this, Italian physician Filippo Pacini (1812–1883) had identified the cholera bacterium. Snow had no way of knowing this, and the scientific world generally remained ignorant of Pacini's findings and rejected Snow's idea. Snow's second insight was that water consumption was related to the spread of cholera. What turned him into an epidemiologist was the way he collected raw data by meticulously going door to door to locate cholera victims, examining public records of cholera deaths, and mapping the victims' locations relative to the location of specific water sources. He then used his understanding of statistics to analyze these data. His analysis supported his insights, and the science of practical epidemiology was born.

HISTORICAL CONTEXT AND IMPACT

Although Snow provided statistical proof that cholera was caused by drinking water from specific sources, many physicians and public officials did not believe him. The prevailing medical theory, known as the miasma theory of disease, stated that disease came from bad air. In addition, many members of the clergy and their congregations believed that disease was sent by God as punishment for poor morals. Snow made little headway against these beliefs during his lifetime.

In 1890, German bacteriologist Robert Koch (1843–1910) published what are now known as Koch's postulates. These are four statements that form the basis for the germ theory of disease, without which there would be no science of epidemiology. They declare that the organism causing a disease must be found in all individuals who have the disease. The organism can be isolated and cultured. When introduced into a healthy individual, the cultured organism will cause the disease, and when re-isolated from that individual, the organism will be identical to the one that caused the initial disease. The gradual acceptance that living organisms—not dirt, not bad odors, and not bad morals that anger God—cause disease allowed scientists to make advances in the science of epidemiology.

In 1951, the United States recognized the importance of epidemiology when it established the Epidemic Intelligence Service (EIS) to guard against biological warfare during the Korean War. Today, the EIS operates under the Centers for Disease Control and Prevention and partners with state and local governments to investigate infectious disease outbreaks such as those that have occurred due to contaminated salad greens. It also responds to public health threats brought on by natural disasters anywhere in the world. On a continuous basis, the EIS epidemiologists conduct surveillance of established and emerging diseases such as monkeypox and Ebola virus disease in order to better understand them and prevent their spread. Infectious diseases are not the only target of the EIS. Their epidemiologists also study chronic diseases, environmental and occupational health threats, and patterns of injuries, birth defects, and developmental disabilities.

The EIS has played a role in the eradication of smallpox (see entry 52), the near eradication of polio, and the identification of the AIDS virus, in addition to assisting in many state and local outbreaks related to contaminated food and water. In 2020, epidemiologists worldwide played a major role in tracking the coronavirus, understanding how it spreads and mutates, and organizing contact tracing of people known to have been exposed to the virus.

Despite the effectiveness of the EIS, outbreaks of contagious diseases are common. Cholera remains present in many parts of the world. Outbreaks frequently occur after natural disasters such as floods, hurricanes, and earthquakes or in war zones when sewer systems are overburdened or destroyed and raw sewage enters the drinking water supply. The World Health Organization estimates that between 1.4 million and 4 million cases of cholera occur each year.

The development of an oral hydration fluid containing sugar and essential salts has made it possible to survive cholera, but only if that fluid is immediately available to replace water loss. A vaccine is also available against cholera, but it is only 50–60% effective in preventing infection, and unlike many vaccines, its protection is relatively short-lived. What started with John Snow looking for the cause of cholera deaths in 1854 has led to epidemiological studies that have prevented millions of cases of disease and death over the past 150 years.

FURTHER INFORMATION

Extra History. "England: The Broad Street Pump—You Know Nothing, John Snow #1–3." YouTube, November 14, 2015. https://www.youtube.com/watch?v=TLpzHHbFrHY&li st=PLhyKYaOYJ_5A1enWhR5Ll3afdyhokVvLv

Jay, Mike. "Medical London: John Snow and the Cholera Outbreak of 1854 with Mike Jay." YouTube, November 11, 2008. https://www.youtube.com/watch?v=Pq32LB8j2K8

Queijo, Jon. "How Cholera Saved Civilization: The Discovery of Sanitation." In Breakthrough!, 27–44. Upper Saddle River, NJ: Pearson Education, 2010.

18. First Use of Antiseptic Spray, 1865

British surgeon Joseph Lister (1827–1912) was the first person to use an antiseptic spray to prevent infection of wounds and surgical incisions. Lister was born into a devout Quaker family in Upton, England. His father, an amateur scientist, developed a microscope that solved the problem of color distortion of specimens. For this he was elected to the Royal Society, the oldest and most respected scientific organization of the time.

Lister received an education from his father in natural science and use of the microscope. His formal education was completed at two Quaker schools, after which he enrolled in University College, London. Lister was an outstanding student, and in 1853, he graduated with an honors medical degree and became a member of the Royal College of Surgeons. He then moved to Edinburgh to study under James Symes, a prominent surgeon. Eventually, he was appointed surgeon to the Glasgow Royal Infirmary.

The Infirmary had built a new surgical building hoping that it would reduce the large number of deaths by overwhelming infection (sepsis). Unfortunately, the new building changed nothing. Lister reported that between 1861 and 1865, almost half the patients in the male accident ward died of sepsis caused by infected wounds. Concern about the rate of infection led Lister to consider its causes.

In the mid-1860s, most physicians believed that disease could be caused by any number of things such as bad diet, poor morals, or punishment from God, but bad air was the leading candidate. Lister initially believed that tiny dust particles in the air caused infection. He did not conceive of these particles as living organisms. His antiseptic spray was intended to protect wounds from this infection-causing air dust.

Carbolic acid (now called phenol) was an antiseptic used to clean sewers and control their odor. On August 12, 1865, Lister first successfully used a spray of dilute carbolic acid on James Greenlees, an 11-year-old boy who had been run over by a cart. Greenlees had a compound fracture of the lower leg and was taken to the Glasgow Royal Infirmary. A compound fracture is one where the broken bone breaks through the skin. Up to that time, the standard procedure for compound fractures was to amputate the limb because the risk of sepsis resulting in death was by some estimates around 80%. Instead, Lister cleaned the skin with carbolic acid, splinted the bones, and bandaged the wound with cloth soaked in carbolic acid. The carbolic acid–soaked bandages were replaced regularly. In six weeks, Greenlees had healed without any sign of infection.

Lister used this technique on other patients but waited until March 1867 to publish a series of articles in *The Lancet*, a British medical journal, outlining the procedures and results of 11 surgeries in which carbolic acid spray was used. His articles gave specific instructions on how to dilute the carbolic acid and administer it with a sprayer similar to what one might use for perfume.

During the time between the first experiment with what he called "the carbolic method" and publication of his work, Lister read about the fermentation experiments of Louis Pasteur (1822–1895) and came to question the bad air theory of disease. He became persuaded that living organisms in the air that were too small to be seen were what caused infection. He presented his ideas at an annual meeting of the British Medical Association and was met with disbelief and hostility. Most surgeons rejected the germ theory and the need to use carbolic acid as an antiseptic.

The Germans successfully experimented with Lister's carbolic method during the Franco-German War (1870–1871) and were the first to widely accept Lister's ideas. Other countries, especially England, remained resistant. To counter this resistance, on October 26, 1877, Lister performed a public operation on Francis Smith, a patient with a compound fracture of the kneecap, to demonstrate his antiseptic method and skill as a surgeon. The patient fully recovered without infection. Over the next two decades, as proof emerged that microbes caused disease, Lister's antiseptic methods became standard practice for surgeons.

Lister retired in 1893. In 1897, Queen Victoria granted him the title of 1st Baron Lister of Lyme Regis in recognition of his contributions to medicine. Other honors followed, and today Lister is considered the father of antiseptic surgery. In his last years, Lister became both blind and deaf. He died at age 84, having lived to see his ideas about the need for antiseptic surgery universally accepted.

HISTORICAL CONTEXT AND IMPACT

Joseph Lister was not the first doctor to promote the idea of infection control. Ignaz Semmelweis (1818–1865), a Hungarian-born physician, attended medical school in Vienna, Austria. Upon graduation in 1847, he received an appointment in obstetrics at a teaching hospital in that city. The hospital had two identical maternity wards, one for training doctors, the other for training midwives. Semmelweis observed that the mortality rate from puerperal fever, better known as childbed fever, was between 13% and 18% in the doctors' training ward while it was about 2% in the midwives' training ward. After eliminating conditions that were similar in both wards, Semmelweis realized that the difference was that doctors did autopsies while midwives did not. The doctors often went directly from performing autopsies to delivering babies without washing their hands or instruments. Semmelweis formulated the rather gruesome idea that the doctors were transferring infection from corpses to women in labor.

This idea was considered ridiculous by most doctors. The germ theory of disease had not yet been developed, and the doctors scoffed at the idea that some unknown, unseen particles could cause infection. They also resented the implication that they—the highly esteemed professionals—were unclean. Despite this, Semmelweis required that doctors wash their hands in a chloride lime solution and the death rate from infection in the doctors' ward dropped to 2%, equivalent to the midwives' ward. Later, when he insisted that instruments be washed, the death rate fell to about 1%.

Despite his success, Semmelweis was not reappointed to his position. He returned to Hungary where he accepted a minor medical appointment. Eventually, he had a mental health crisis and began writing letters to doctors who would not wash their hands calling them mass murderers. He was committed to a mental institution and died 14 days later of infection in a wound that he acquired from fighting with the guards.

Fortunately, theory eventually caught up with observational findings, and with increased acceptance of the germ theory, antiseptic practices became the norm. The understanding of germs soon spread beyond the medical community. Beginning in the late 1870s, more emphasis was put on cleanliness. In 1879, American chemist Joseph Lawrence developed an alcohol-based mouthwash that he named Listerine in honor of Joseph Lister. By the mid-1880s, cleanliness became associated with respectability and an indication of moral and social standing. Those who were clean were considered to have better morals and more refined sensibilities than those who were dirty.

As hospitals increasingly emphasized cleanliness, their image changed from being institutions of last resort to places of healing. In 1885, Ernst von Bergmann (1836–1907) introduced the use of the autoclave, invented in 1879 by microbiologist Charles Chamberland (1851–1908), to steam-sterilize surgical instruments and dressings. In 1898, American William Stuart Halstead (1852–1922) introduced the use of sterile rubber gloves during surgery. The first gloves for medical use were made by the Goodyear Rubber Company and had to be sterilized by the user. Presterilized gloves did not become available until the 1960s.

These developments, along with prohibiting surgeons from wearing street clothes while operating and setting aside dedicated rooms for operations, marked a change in emphasis from antiseptic procedures to asepsis. Antiseptic methods kill microbes that are present. Aseptic methods aim to prevent the introduction of any microbes into the environment. The need for clean, controlled conditions carried over into the manufacturing of pharmaceutical drugs, resulting in the 1906 Pure Food and Drug Act (see entry 28).

Today a combination of rigorous aseptic and antiseptic methods is used in all surgeries. Despite this, surgical site infections do occur in between 2% and 4% of inpatient hospital surgeries in the United States. Although many surgical infections can be cured with antibiotics, about 3% of those who become infected, or about 20,000 people each year, will die. With the increase in antibiotic-resistant microbes, there is concern that this number will increase; thus, hospitals have become even more concerned with providing a germ-free environment for surgery.

FURTHER INFORMATION

Best, M., and D. Neuhauser. "Ignaz Semmelweis and the Birth of Infection Control." *Quality and Safety in Health Care* 13, no. 3 (June 2004): 233–234. https://qualitysafety. bmj.com/content/13/3/233.full

History Pod. "12th August 1865: Joseph Lister Carries Out World's First Antiseptic Surgery." YouTube, August 11, 2015. https://www.youtube.com/watch?v=ZOYA00mE4MQ

Pitt, Dennis, and Jean-Michel Aubin. "Joseph Lister: Father of Modern Surgery." *Canadian Journal of Surgery* 55, no. 5 (October 2012): E8–E9. https://www.ncbi.nlm.nih.gov /pmc/articles/PMC3468637

Worboys, Michael. "Joseph Lister and the Performance of Antiseptic Surgery." *Notes and Records: The Royal Society Journal of the History of Science* 67, no. 3 (September 20, 2013): 199–209. https://www.ncbi.nlm.nih.gov/pmc/articles/PMC3744349

19. First Successful Brain Tumor Operation, 1879

The first successful operation to remove a brain tumor occurred on July 21, 1879 in Glasgow, Scotland, although this claim has been disputed. On that date, surgeon William Macewen (1848–1924) removed a brain tumor from 14-year-old Barbara Watson. Watson had initially developed a lump above her eye which Macewen had removed a year earlier. The lump, however, had returned. Macewen intended to repeat the removal process, but in addition to the lump regrowing, Watson was now having seizures. Seizures made Macewen suspect a brain tumor. By documenting the parts of the body that were affected by the seizures, he was able to determine the location of the suspected tumor in the brain.

Using chloroform to anesthetize Watson, Macewen made a hole in the skull. His diagnosis was correct. The lump over her eye was a tumor that had grown through the skull and into the brain.

Three protective layers called meninges cover the brain. The outermost layer, the dura mater, is a tough fibrous layer between the skull and the softer brain tissue. Watson's tumor had spread across the dura mater. Macewen successfully removed it. The girl recovered but died a few years later from kidney failure.

William Macewen received his medical degree from the University of Glasgow in 1872 and began his surgical career at the Royal Glasgow Infirmary where he trained under Joseph Lister (1827–1912), the father of infection control (see entry 18). He was so strongly influenced by Lister's ideas about asepsis and wound infection that he started a nurse training program that emphasized handwashing, use of antiseptic sprays, and sterilization of surgical instruments, none of which were common practices at the time.

After Watson's operation, Macewen described the procedure in the 1879 *Glasgow Medical Journal*. He even took Watson to a meeting of the Glasgow Pathological and Clinical Society as proof of his success, but the procedure received little professional attention and was ignored by the popular press. Macewen performed six more brain surgeries before Rickman Godlee (1849–1925), a well-connected London surgeon, claimed that he had performed the first successful brain tumor removal on November 25, 1884.

Rickman Godlee was raised in a prosperous, well-connected Quaker home. His mother was Joseph Lister's sister. After receiving a medical degree at the University of London, Godlee went to Edinburgh where he worked as Lister's personal assistant before returning to London to practice surgery. On November 25, 1884, at the Hospital for Epilepsy and Paralysis in London, Godlee performed what he considered to be the first brain tumor removal operation. The patient, a farmer named Henderson, had been diagnosed with a brain tumor by neurologist Alexander Hughes Bennett (1848–1901). Godlee removed a tumor the size of a walnut. Henderson's symptoms—headaches and seizures—disappeared. Godlee considered the operation a success, but Henderson died from infection 28 days later.

Godlee's operation caused a sensation in the popular press. The *London Times* ran a prominent article calling it the world's first brain operation. The article also ran in the *Glasgow Herald*. Immediately, a flurry of outraged letters from Glasgow descended on editors of both newspapers and several medical journals. The Glasgow physicians protested that one of their own had been performing brain tumor removal for several years. They claimed that these surgeries had been ignored by snobbish London physicians who considered London the leader in all important British medical innovations. The London physicians responded saying that Macewen did not perform a true brain operation because he only removed a tumor from the dura mater and not from deeper underlying brain tissue. They found it irrelevant that Macewen's patient survived while Godlee's did not.

Eventually, the *British Medical Journal* grudgingly agreed that Macewen had, indeed, performed the first brain tumor removal, but they made a distinction. Macewen, they said, removed a secondary tumor—that is, one that began elsewhere and migrated into the brain. Godlee, on the other hand, was the first person to remove a primary brain tumor—one that originated in the brain.

Macewen was knighted in 1902 and went on to help found the Princess Louise Scottish Hospital for Limbless Sailors and Soldiers (now Erskine Hospital) where he worked with engineers to design artificial limbs for those who had been injured in World War I. Godlee went on to serve the royal family as surgeon to the household of Queen Victoria and surgeon to Edward VII and George V. Godlee was named a baronet in 1912 and wrote a well-received biography of his famous uncle, Joseph Lister.

HISTORICAL CONTEXT AND IMPACT

Today we accept as established fact that different parts of the brain regulate different functions, but in the 1870s when Macewen and Godley were practicing medicine, there were three competing theories of the brain. The phrenology theory claimed that the size, shape, and irregularities of the skull determined personality and mental ability. The unitary brain theory promoted the idea of the brain as a homogenous organ in which all parts functioned in the same way. The third theory, which was gaining popularity by the mid-1870s, was known as the localization theory. It hypothesized that different parts of the brain controlled different actions.

Neither Macewen's operations nor Godlee's would have succeeded without work to support the localization theory. This work was done by two physicians, John Hughling Jackson (1835–1911) and David Ferrier (1843–1928). John H. Jackson was a physician at the National Hospital for Paralysis and Epilepsy (now the National Hospital for Neurology and Neurosurgery) in London. Through careful observation of patients' symptoms, behaviors, and autopsy results, he related certain behaviors (such as seizures) or deficits (such as loss of ability to speak or paralysis of a muscle) to damage in specific and consistent parts of the brain.

David Ferrier (1843–1928), a Scottish neurologist and friend of Jackson, conducted experiments in which he electrically stimulated specific areas of the brains of dogs and monkeys. He showed that electrical stimulation of the same area of the brain always produced a corresponding muscle movement. This supported the localization theory.

Ferrier mapped the different areas of the cortex—the brain's outermost layer—of monkey brains and correlated the areas with specific physical signs and symptoms. He also hypothesized that the monkey brain was similar enough to the human brain for it to be used to correlate symptoms such as seizures to specific human brain locations. Both Macewen and Godlee were familiar with Ferrier's work and used his brain map to successfully locate and remove their patients' brain tumors. Ferrier, however, was accused by anti-vivisectionists of cruelty to animals and in 1881 was the first scientist to be prosecuted under the 1876 Cruelty to Animals Act. He later went on to found the National Society for Epilepsy along with Jackson and William Gowers (1845–1915), another skilled neurologist.

The use of brain localization maps in surgery spread quickly. In 1888, surgeon William W. Keen (1837–1932) became the first American to successfully remove a brain tumor. His patient lived another 30 years. Beginning in 1895, diagnostic X-rays (see entry 25) gave surgeons a way to see inside the brain before surgery and solidified the localization of brain function theory. More recent developments such as computed tomography (see entry 47) and magnetic resonance imaging (see entry 50) give surgeons even more information about brain tumors before they attempt their removal.

The first robot-assisted surgery, which took place in 1985 (see entry 53), was done in order to biopsy a brain tumor and determine if it was malignant. Modern neurosurgeons sometime insert a thin laser probe into the brain to destroy tumors, something that could not be done without brain mapping. Brain maps have also improved. Transcranial magnetic stimulation (TMS) has eliminated the need to stimulate the brain with an electric current. With TMS, researchers can safely stimulate specific areas of the brain and ask patients what they see, hear, smell, feel, or remember during stimulation.

Today there are about 3,500 board-certified neurosurgeons in the United States. These surgeons operate on tumors of the brain and spinal cord. Although the chance of developing a primary cancerous brain tumor during one's lifetime is less than 1% and is even lower for a spinal cord tumor, each year about 25,000 American adults and 4,200 children under age 15 receive these diagnoses. Secondary tumors that originate somewhere else in the body and migrate to the brain are more common. Overall, brain and nervous system tumors were the tenth leading cause of death in the United States as of 2020. Individual survival rates vary significantly with age and type of tumor but have improved significantly in the twenty-first century.

FURTHER INFORMATION

McArdle, Helen. "William Macewen: Meet the Pioneering Glasgow Surgeon Behind the World's First Brain Tumor Op." *The Herald*, May 25, 2019. https://www.heraldscotland.com/news/17660679.william-macewen-meet-pioneering-glasgow-surgeon-behind-worlds-first-brain-tumour-op

Neuroscientifically Challenged. "2-Minute Neuroscience: Brain Tumors." YouTube, September 20, 2018. https://www.youtube.com/watch?v=pBSncknENRc

Royal College of Physicians and Surgeons of Glasgow. "World's First Brain Tumour Removal." YouTube, December 10, 2019. https://www.youtube.com/watch?v=pjSV_gsXME0

Shreykumar, P. S., K. P. Biren, R. H. Darshan, C. V. George, and V. H. Easwer. "History of Brain Tumor Surgery—A Global and Indian Perspective." *Archives of Medicine and Health Sciences* 9 (2021): 156–162. https://www.amhsjournal.org/article.asp?issn=2321-4848;year=2021;volume=9;issue=1;spage=156;epage=162;aulast=Shreykumar

20. First Baby Incubator, 1880

The first baby incubator was installed in the Paris Maternity Hospital in 1880. In the 1880s, about one-quarter of all children died before their first birthday. Babies born prematurely—that is three or more weeks before a full-term pregnancy of 40 weeks—had almost no chance of surviving. Premature babies usually have serious breathing problems because a protein needed for the lungs to fully expand is not made in adequate quantities until the last few weeks before birth. In addition, premature babies have difficulty maintaining their body temperature and their immature immune system makes them more vulnerable to infection.

Stéphane Tarnier (1828–1897), the son of a country doctor, studied medicine in Paris, receiving his medical degree in 1850. While obtaining additional training as an obstetrician, he became interested in the period just before and after birth and even invented a special kind of forceps to be used during delivery. In 1878, Tarnier visited the Paris Zoo. There he saw eggs of exotic birds being hatched in a warm, enclosed incubator. It struck him that a similar device might help premature babies survive by providing them with a warm, stable, sheltered environment. He asked Odile Martin, a zoo employee, if he could make an incubator large enough for a baby.

Not much is known about Odile Martin except that he was from Neuilly, a village west of Paris. He is variously described as the zoo's poultry raiser or the zoo's instrument maker. He did, however, make a baby incubator for Stéphane Tarnier. This was the first closed incubator for hospital use.

Martin's incubator was large enough to hold four babies. It had a double wall with the space between the walls filled with sawdust for insulation. The lid was made of thick double-glazed glass. The incubator sat on a heated water tank. A thermosiphon—a device that uses natural forces generated by heat transfer to passively circulate air or water without a pump—introduced warm air through vents at the bottom of the incubator. Air exited the incubator through vents at the top. One problem with this incubator was temperature control. Without constant vigilance, it could overheat and kill the babies. An early modification was making a smaller, easier-to-move incubator that would hold a single baby. By 1883, the survival rate of premature babies weighing less than 4.4 pounds (2,000 g) who were cared for in an incubator went from 35% to 62%.

HISTORICAL CONTEXT AND IMPACT

Stéphane Tarnier was not the first person to experiment with ways to keep premature babies warm. In 1857, Jean-Louis-Paul Denucé (1824–1889) of Bordeaux, France, described a warming cradle that consisted of a double-walled zinc tub. The space between the walls was filled with warm water. Carl Credé (1819–1892), a German obstetrician and gynecologist of French heritage, built a similar double-walled warming crib in 1864, although his main interest was in preventing blindness in babies born to mothers infected with gonorrhea. Both these warming cradles, unlike Tarnier's incubator, were open to the air. It is unclear how effective they were.

Within a few years, improvements were added to the basic incubator. A thermometer was added, but the heat level still had to be monitored frequently by attendants. Pierre Budin (1846–1907), who succeeded Tarnier as head of obstetrics at the Paris Maternity Hospital, added a thermostat that was connected to an alarm that would sound if the temperature became too high. He also added filters to help clean the air entering the incubator and developed a human milk feeding program using special spoons and tubes for babies too premature to breastfeed.

By defining the factors that threatened premature babies—such as temperature, disease, and inadequate diet—and demonstrating ways to overcome these threats, he established the beginnings of the field of neonatology. Nevertheless, this specialty was not recognized by name until 1960. The Budin incubator, with minor changes, was duplicated in the United States by physician John Bartlett and first used in a hospital in Chicago in 1887.

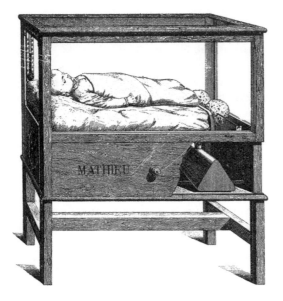

Watching baby birds hatch in an incubator at the Paris zoo inspired the construction of the first baby incubator. Before baby incubators almost all premature babies died. (Artificial incubator for premature babies, by Mathieu. Wellcome Collection. Attribution 4.0 International (CC BY 4.0))

Incubators were expensive for hospitals to use because they required round-the-clock attention from nurses. Then in 1889, Alexandre Lion, a physician in the south of France, patented a cabinet-style incubator. His incubator needed fewer staff to operate, thus reducing the cost to hospitals. But premature babies still needed extra attention and had to be fed every two hours, so the cost of premature baby care, which sometimes lasted for months, remained out of reach for most parents. To combat this, Lion established a total of seven incubator charities in France and Belgium between 1894 and 1898. He installed his incubators in store fronts, at fairs, and at expositions and then charged an admission fee for the public to see the babies. Lion's charities accepted premature babies regardless of the parents' ability to pay and used the admission fees and donations to pay for their care. The charities treated about 8,000 premature babies during the four years that they existed and claimed to have an outstanding 72% survival rate.

Pierre Budin found that many in the medical community were reluctant to use incubators, so he developed a program similar to that of Dr. Lion. Budin sent Dr. Martin Couney (1869–1950) to the Berlin Exposition of 1896 where he set up what he called a "child hatchery" of six premature incubator babies and charged the public to see the babies. The program was a financial success. Over the next 40 years, Couney was shunned by hospitals and the medical community, so he set up baby incubators at world fairs in Paris, Buffalo, Chicago, and New York. However, his most successful and long-lasting venture was at the amusement park at Coney Island, New York.

In the early 1900s, Coney Island's amusement area was the largest in the world, attracting several million visitors each year. In 1903, Couney set up a baby incubator at Luna Park, among curiosities such as the bearded lady and the sword swallower. This exhibit attracted millions of viewers at 25 cents each and continued until 1943. A second exhibit at Coney Island's Dreamland existed from 1904 until Dreamland burned down in 1911.

The admission fees from these exhibits were used to pay registered nurses to care for the babies, as well as a support staff of cooks, wet nurses, exhibit guides, and chauffeurs to take workers home after their shifts. Parents paid nothing for the round-the-clock care. There are stories of parents rushing their babies to Coney Island for placement in the exhibit after a hospital had told them that there was no hope for their child to survive or when the parents could not afford hospital care. About 8,000 babies were cared for in the Coney Island exhibits, of which about 6,500 survived. National Public Radio interviewed one of these survivors in 2015 (see Further Information following this entry). Couney set up a similar, but smaller, baby incubator exhibit on the boardwalk at Atlantic City, New Jersey, from 1902 to 1943. By the early 1950s, baby incubators had been embraced by hospitals and Couney's exhibits were discontinued.

Incubators basically remained unchanged from the early 1900s to the early 1940s, when physician Charles Chapple (1904–1979) invented an incubator called the isolette which has since been updated and improved. Between 1950 and 1970, hospitals gradually established newborn intensive care units (NICUs). In 1975, the American Board of Pediatrics recognized neonatology as a subspeciality of pediatrics. Today NICUs and neonatologists can be found in most major hospitals.

Survival of premature babies has improved dramatically in the past half century. A 2.2-pound (1 kg) baby born in 1960 had only a 5% chance of survival. A baby of the same size in 2000 had a 95% chance of survival. The smallest baby to survive, identified only as Baby Saybie, was born by cesarean section in December 2018 at 23 weeks, after her mother suffered life-threatening complications from the pregnancy. At birth, Baby Saybie was 9 inches (23 cm) long and weighed 8.3 ounces (235 g). She spent five months in the NICU before being declared healthy enough to go home. She is an extreme exception in beating the odds of survival. As of 2020, only 17% of babies born at 23 weeks survive. This increases to 50% at 25 weeks and 90% at 27 weeks. Survival rates do not tell the entire story. Extremely premature babies commonly have more lingering health and developmental problems than full-term babies.

The ability to keep extremely premature babies alive has an impact on the abortion debate in societies where abortion is legal. Decisions must be made about when in pregnancy the fetus can potentially survive outside the womb and when abortion should no longer be an option. This is an issue that is likely to remain contentious in the future if the survival rate of extremely premature babies continues to improve.

FURTHER INFORMATION

"Babies On Display: When a Hospital Couldn't Save Them, a Sideshow Did." National Public Radio, July 10, 2015. https://www.npr.org/2015/07/10/421239869/babies-on-display-when-a-hospital-couldnt-save-them-a-sideshow-did
Rebovich, Kelsey. "The Infant Incubator in Europe (1860–1890)." The Embryo Project, February 11, 2017. https://embryo.asu.edu/pages/infant-incubator-europe-1860-1890

Reedy, Elizabeth."Care of Premature Infants." University of Pennsylvania Nursing, Undated. https://www.nursing.upenn.edu/nhhc/nurses-institutions-caring/care-of -premature-infants

Stern, Liz. "Blast from the Past: Tiny Incubator Babies: The Coney Island Attrac- tion." New York Historical Society, August 1, 2014. https://www.nyhistory.org/blogs /blast-past-tiny-incubator-babies-coney-island-attraction

21. First Steps toward Universal Health Care, 1883

The World Health Organization (WHO) defines Universal Health Coverage (UHC) as: "[A]ll people have access to the health services they need, when and where they need them, without financial hardship. It includes the full range of essential health services, from health promotion to prevention, treatment, reha- bilitation, and palliative care."

The first mandated health insurance system was the Sickness Insurance Law and Social Health Insurance Program (SHI) set up in 1883 by German chancel- lor Otto von Bismarck (1815–1898). This law required employers and employees in high-risk occupations, mostly industrial workers in rail, ship, and metal works, to contribute to a health insurance fund.

Bismarck set up the SHI system at least partly for political reasons. The oppo- sition Social Democratic Party was heavily supported by industrial workers, the same workers who would be covered by the SHI. The introduction of the SHI helped shift their vote to keep Bismarck's party and the monarchy it supported in power in the newly unified Germany.

This German insurance plan did not include all medical coverage, but it did cover sick pay, drugs, and death benefits. It could not be considered univer- sal since only 10% of workers were covered. More workers were added in later years, including government workers in 1914, the unemployed in 1918, retirees in 1941, students in 1975, and artists in 1981. It did not become truly universal until 1988, over 100 years after Bismarck initiated the program.

The WHO states that for a system to meet its definition of universal health care, it must be affordable to all. A system where funding comes from a combina- tion of businesses and the insured will force participants to pay higher premiums if the cost of health care increases. To maintain affordability, the German system originated by Bismarck is now subsidized by the government. As of 2018, German UHC was about 85% government-funded and Germany spent just over 11% of its gross domestic product (GDP) on health care, one of the highest percentages in Europe.

HISTORICAL CONTEXT AND IMPACT

A state-sponsored healthcare system existed thousands of years before that of Germany. Texts discovered during an excavation of the ancient Egyptian city of Deir el-Medina, occupied between 1550 and 1070 BCE, gave information on skilled laborers' absences from work and payments made to them in the form of grain allotments. The texts show that workers off work for several days due to illness got full grain allotments. The sick workers were also provided with a doctor and his assistant, paid for by the state. This was certainly not UHC since only the most skilled workers benefited—but it shows that health care and sick days were considered important even millennia ago.

In the late Roman Republic and early Roman Empire, an Aesculapium, or temple to the god of healing, was established. There is no record of fees being collected for a stay at an Aesculapium. The funding model was that of any temple, with contributions based on certain conditions such as successful healing and the ability to pay. Funds for building the Aesculapium came from the state treasury or from taxes.

The roots of UHC in Europe date back to the 1300s. Guilds of workers established mutual assistance funds for healthcare costs that included both payments to doctors and payments for those unable to work. These were limited to guild members, who made up only about 5% of the population. The rest of the population depended on charitable organizations, often religious, that ran hospitals.

In the late eighteenth century, states in Europe began to restrict the power of guilds in order to create a more liberal labor market open to all. This led to the banning of guilds in revolutionary France and later in other countries. Trade associations stayed strong in Germany and formed the foundation of Bismarck's SHI system.

The first instance of state involvement in health care occurred in the late nineteenth century in Sweden and Norway. Swedish district physicians were given royal commissions if they agreed to treat the poor free of charge. Hospitals had a different funding mechanism. In the first part of the nineteenth century, hospitals in England were primarily funded by charities, sponsored either by the nobility or by the church and later by businesses. Before sanitary practices were established, the middle and upper classes stayed away from hospitals. They were treated at home and paid the provider directly for the service. Hospitals were for the working class and the poor. A person called an almoner interviewed incoming patients to determine how much, if anything, that person would pay.

As hospitals became better at providing effective treatment, the middle class began to use them. They were willing to pay for care, and in return, the hospitals began providing private rooms for them instead of the large wards where patients who could not afford to pay were treated. As more patients could afford to pay, the perceived need for charitable funding decreased.

Beginning in 1938, New Zealand set up a healthcare insurance system that was much closer to the modern conception of UHC. The Labour government, which first came to power in 1935, had UHC funded by the state as part of its platform. Its first proposal was to pay doctors a flat fee per patient that would cover all visits. Doctors objected to this since they felt it made them effectively state employees and not independent professionals. They preferred a fee-for-service system in which they would be paid by the government for each patient visit.

In 1939, women were provided free maternity care and general practitioners were given free access to hospitals, but it took until 1941 to completely implement a government-paid fee-for-service system. Under this system, doctors were permitted to charge patients additional fees, but when the system first began, few did.

The New Zealand system was closer to being universal than the original German system. It also had more of its costs paid for by government as opposed to the German system where employee and employer payments were supplemented by government funding. As of 2017, New Zealand spent 9% of its GDP on health care. It has one of the lowest levels of medication use among developed countries.

UHC programs that guarantee health care for all began to spread after World War II. Belgium adopted UHC in 1945, the United Kingdom established the National Health Service in 1948, South Korea adopted a UHC program in 1988, and Japan in 1972. In the United States, charity hospitals began to be replaced by private hospitals funded by patient fees. By the beginning of the twentieth century, the average American needed some sort of health insurance to pay for hospital care, but UHC was not adopted. During World War II, government-mandated wage caps in the United States caused employers to offer employees paid health insurance as a way to retain scarce workers and provide additional compensation outside the wage cap. The closest the United States has come to UHC is with the Medicare program begun in 1965. For most people over age 65, the federal government will pay for a large portion of their medical expenses.

Not all countries have equal levels of UHC. WHO says that progress toward UHC can be measured by (1) the proportion of a population that can access essential quality health services and (2) the proportion of the population that spends a large amount of household income on health care. The initial German system was not effective by the first criterion. The initial New Zealand system paid doctors enough so that co-pays were low or zero, but as time went on doctors' fees increased faster than government payments for them. Doctors then legally began adding supplemental charges to be paid by the individual, so the system became less effective in meeting the second criterion.

Funding for UHC comes from a mixture of public and private sources. In compulsory health insurance models (e.g., Germany), each citizen must purchase health insurance, although the premiums are subsidized by the government. In single-payer plans, the government pays healthcare providers directly from tax revenue. Health services in single-payer plans can be private, as in Canada, or publicly owned as in Great Britain.

While a substantial number of countries have implemented UHC since 1883, there are still many in the world for whom health care is a substantial expense.

The World Bank Group in a 2019 analysis reported that 925 million people worldwide spend more than 10% of their household income on health care while over 200 million spend over 25%. The percentage of people spending more than 10% of their income on health care ranges from 17.2% in South Asia to 4% in North America.

FURTHER INFORMATION

Belgrave, Michael. "Primary Health Care—Improving Access to Health Care, 1900s–1970s." Te Ara—The Encyclopedia of New Zealand, Undated. https://teara.govt .nz/en/primary-health-care/page-3

Bump, Jesse B. "The Long Road to Universal Health Coverage: Historical Analysis of Early Decisions in Germany, the United Kingdom, and the United States." *Health Systems and Reform* 1, no. 1 (2015): 28–38. https://www.tandfonline.com/doi/full/10.4161 /23288604.2014.991211

Clarke, Laura. "Some Ancient Egyptians had State-Sponsored Healthcare." *Smithsonian Magazine*, February 20, 2015. https://www.smithsonianmag.com/smart-news/some -ancient-egyptians-had-state-sponsored-healthcare-180954361

World Health Organization. "Universal Health Coverage." Undated. https://www.who .int/health-topics/universal-health-coverage#tab=tab_1

22. First Cancer Immunotherapy Treatment, 1891

Immunotherapy is the introduction of a biological substance into the body that stimulates the immune system to fight cancer cells. In 1891, William Bradley Coley (1852–1936), a physician in New York City, was the first person to use an immunotherapy technique to treat cancer, even though he did not understand the science that made his procedure work. Today, Coley is recognized for his pioneering research, but during his lifetime, the medical community widely rejected his treatment method. Not until the late 1970s did scientific interest in immunotherapy as a cancer treatment reemerge.

William Coley was born into a privileged Connecticut family. He attended Yale University and received a medical degree from Harvard in 1888, after which he worked as a surgeon at New York Memorial Hospital (now part of Memorial Sloan-Kettering Cancer Center). His area of specialization was bone and soft tissue cancers called sarcomas, and he became head of the hospital's Bone Tumor Service.

Coley's interest in cancer began in 1890 when he met 17-year-old Bessie Dashiell who had an aggressive bone tumor in her hand. Unable to surgically remove the tumor, Coley's only option was to amputate Bessie's forearm. The cancer, however, had already spread, and Bessie died ten weeks later despite the amputation. Her death stimulated Coley to look for a better way to treat inoperable cancers.

Coley began his search by looking at hospital records of former patients diagnosed with sarcomas. He found a patient named Fred Stein who seven years earlier had an inoperable neck tumor that disappeared after he developed a case of erysipelas. Erysipelas is a serious bacterial skin infection caused by *Streptococcus pyogenes*. Coley was so intrigued that he searched for Stein and eventually found him living cancer-free in a Lower Manhattan tenement. Through an extended search, Coley located 47 "miraculous cure" cases where tumors had disappeared after the patient developed a bacterial infection.

In 1891, soon after Coley made the connection between streptococcal infection and tumor remission, he met an Italian immigrant named Zola who had an inoperable tumor on his tonsil large enough to interfere with his ability to eat and talk. Since Zola's tumor would eventually kill him, Coley tried injecting streptococcal bacteria directly into the tumor. This produced a small reduction in tumor size but not enough to eliminate the tumor. Coley kept trying and eventually injected Zola with a more potent strain of streptococcus. Zola developed a serious streptococcal infection, but the tumor liquified and disintegrated.

Coley continued to treat patients who had inoperable tumors with *S. pyogenes*. Of the first 10, two died as a result of strep infection, but all, even the patients who died, had their tumors shrink or completely disappear. Coley was convinced that infection and the fever it caused were essential to tumor remission, but because of the risk of death from streptococcal infections in the pre-antibiotic era, he looked for less risky ways of producing tumor shrinkage. Eventually, he settled on a combination of heat-killed *S. pyogenes* and another bacterium, *Serratia marcescens*. This combination became known as Coley's Toxin. Coley had created the first cancer vaccine through observation, trial, and error.

When Coley published the results of this treatment, his claims were met with skepticism to the degree that some doctors called him a charlatan. As early as 1894, the *Journal of the American Medical Association* published an editorial questioning the credibility of Coley's work, claiming that it was a failure and that there was no future in Coley's Toxin as a cancer treatment. Many doctors, the journal said, had been unable to reproduce Coley's results. This was likely true because Coley's Toxin was difficult to make and varied in strength from laboratory to laboratory. In addition, the dose had to be individualized to produce a sustained fever, something Coley thought was essential to the success of the treatment. Some of Coley's patients had other cancer treatments in addition to Coley's Toxin, potentially compromising Coley's conclusions and leading to charges that his experiments were poorly documented.

Despite the poor reception to his toxin, Coley continued to inoculate patients with inoperable tumors, eventually treating almost 1,000 people with his mixture. His documented success rate in achieving remission was about 10%. The pharmaceutical company Parke-Davis was interested enough in his results to manufacture Coley's Toxin commercially from 1899 until 1952. Despite the skepticism of other doctors, Coley never gave up on his belief in the treatment. In 1902, he received a large grant from the wealthy Huntingdon family to pursue cancer research. This was the first endowment in the United States specifically designated for cancer research, but Coley's Toxin was soon overshadowed by new, potentially more promising cancer treatments.

HISTORICAL CONTEXT AND IMPACT

William Coley was not the first doctor to notice a connection between bacterial infection and tumor remission. Friedrich Fehleisen (1854–1924), a German authority on streptococcal bacteria, wrote a paper documenting five patients whose malignant tumors disappeared after they developed erysipelas. Fehleisen's interest, however, was in bacterial diseases and not cancer, so he did not follow up on his observation.

Coley diligently pursued the relationship between streptococcal infections and tumor remission, but his work was frequently discounted during his lifetime. One problem was that Coley could not explain how his treatment worked. In the early 1900s, the immune system was still a mystery. Today researchers understand that cells have specific proteins on their surfaces. During development, the immune system learns to identify which cells belong to the body and does not attack them. It does attack foreign cells such as bacteria and mutated (altered) body cells such as cancer cells. Often the immune system can prevent illness by eliminating these foreign cells, but sometimes it is overwhelmed, and we get sick. Although Coley did not know it, the bacteria he injected into cancer patients somehow activated the immune system to attack not just the injected bacteria but also the patient's tumor cells. Fever, which Coley thought was so important, was not therapeutic. It was a side effect of the work the immune system was doing.

Coley was thwarted in other ways that he could not control. X-rays were discovered in 1895 and one year later were being used to treat cancer. Treatment with radium, discovered in 1898, soon followed. Unlike Coley's Toxin, radium treatment produced reproducible tumor shrinkage and prompt relief from pain. At the time, the negative side effects of radiation were not known.

Coley also experienced the misfortune of working under Dr. James Ewing (1866–1943), the hospital medical director. Ewing was a strong supporter of radiation therapy and believed that all other methods for treating cancer were inferior. He solicited a $100,000 donation (equal to about $3 million in 2022) from Canadian businessman James Douglas (1837–1918) specifically to further the use of radium as a cancer treatment. The combination of surgery, when possible, and radiation therapy became the predominant method of treating cancer.

Although the American Medical Association revised its position on Coley's Toxin in 1935, saying that it might play a role in cancer treatment, the toxin was further marginalized when chemotherapy came into use in the 1940s. The death blow to Coley's Toxin, however, came in 1962 with the Kefauver-Harris Amendment to the Federal Food, Drug, and Cosmetic Act. The amendment required that before drugs could be sold in the United States, they had to show proof that they were both safe and effective, which was something not previously required. Some drugs, such as aspirin, were grandfathered in and did not require additional testing, but Coley's Toxin was classified as a new drug even though it had been used for 70 years. To be legal, Coley's Toxin would have to go through rigorous clinical trials, and there was no commercial interest in making that investment. The toxin, however, was not forgotten. Based on her father's work, Coley's daughter, Helen Coley Nauts (1907–2001), established the Cancer Research Institute in 1953. Today the Institute is one of the leading research organizations dedicated to advancing cancer immunotherapy.

As researchers began to better understand the immune system, interest in the potential of cancer immunotherapy increased. This resulted in the development of new approaches to cancer treatment. The first monoclonal antibody treatment for cancer was approved by the Food and Drug Administration in 1997 for the treatment of non-Hodgkin's lymphoma (see entry 48). Monoclonal antibodies are laboratory-grown immune system proteins that are engineered to attack specific kinds of cancer cells when injected into the body. In 2010, the first modern cancer vaccine was approved for use in prolonging the life of men with advanced prostate cancer. Today, researchers are looking at a variety of ways in which the immune system and its products can be used or altered to disrupt or destroy cancer cells, and William Coley is now recognized as the father of cancer immunotherapy.

FURTHER INFORMATION

Davis, Rebecca. "Training the Immune System to Fight Cancer Has 19th-Century Roots." National Public Radio Morning Edition, December 28, 2015. https://www.npr.org /sections/health-shots/2015/12/28/459218765/cutting-edge-cancer-treatment -has-its-roots-in-19th-century-medicine

McCarthy, Edward F. "The Toxins of William B. Coley and the Treatment of Bone and Soft-Tissue Sarcomas." *Iowa Orthopaedic Journal* 26 (2006): 154–158. https://www .ncbi.nlm.nih.gov/pmc/articles/PMC1888599

Vazquez-Abad, Maria-Dolores. "Immune Response: Short History of Immuno Oncology Beginnings." YouTube, August 6, 2018. https://www.youtube.com/watch?v=P-uBwh1L7nc

23. First Successful Open Heart Surgery, 1893

Daniel Hale Williams (1856–1931) performed the first successful open heart surgery. On July 10, 1893, James Cornish, a young Black man, went into shock after being stabbed in the chest during a knife-fight. He was taken to the Provident Hospital in Chicago, Illinois. Williams, a Black physician, had founded the interracial hospital in 1891. To save Cornish, Williams performed a risky operation. He exposed the breastbone, cut through a rib, made a small opening to the heart, repaired a damaged internal mammary artery, and closed a hole in the pericardium, the sac surrounding the heart. Fifty-one days later, Cornish left the hospital a healthy man. He lived for at least another 20 years.

One of eight siblings, Daniel Williams, was born in Hollidaysburg, Pennsylvania. His father died when he was nine, and his mother then moved the family several times. Williams was initially apprenticed to a shoemaker in Baltimore, Maryland but disliked the work. Eventually he moved to Janesville, Wisconsin, where he became a successful barber. Ambitious and talented, Williams then became an apprentice under Dr. Henry Palmer (1827–1895), a white man who later became surgeon general of Wisconsin. In 1880, Williams was accepted at the Chicago Medical School (now Northwestern University Medical School) and graduated with a Doctor of Medicine degree.

After graduation, Williams, one of only three Black doctors in Chicago, practiced medicine and surgery at the South Side Dispensary. Over the next few years, he was appointed as doctor to the City Railway Company and the Protestant Orphan Asylum. In 1889, he was appointed to the Illinois Board of Health where he worked on medical standards for hospitals.

In the 1880s, in Chicago and elsewhere, Black doctors were not hired by private hospitals and there were very few opportunities for Black women to receive formal nursing training. Emma Reynolds, sister of Reverend Louis Reynolds, a prominent Black clergyman in Chicago, was one of those women who wanted to become a registered nurse. Unable to enroll in nursing school because of her race, she asked her brother if Williams could help her. Williams was unable to persuade any nursing school to accept Emma. In response, and with the help of many donors including the Armour Meat Packing Company, Williams founded the Black-run but interracial Provident Hospital and Nursing Training School in 1891. At this hospital he performed the first open heart surgery. The Provident Hospital continued to serve patients of all races until it closed in September 1987. In 1993 it reopened in affiliation with the Cook County Bureau of Health Services.

Williams went on to become surgeon-in-chief at Freedmen's Hospital in Washington, D.C., from 1894 to 1898. There he had great success in reducing the mortality rate in surgical cases. In 1895, he helped organize the National Medical Association in response to the American Medical Association's policy of allowing individual states to exclude Black doctors from their state organizations (see entry 13). Williams returned to Chicago in 1902 where he continued to practice medicine. In 1913, he became the first Black doctor to be admitted to the American College of Surgeons. His medical career ended when he had a stroke in 1926. He died five years later.

HISTORICAL CONTEXT AND IMPACT

The operation Daniel Williams performed in 1893 was done under difficult emergency conditions using rudimentary anesthesia. Antibiotic drugs had not yet been developed. Some authorities insist that this operation was not true open heart surgery and call it pericardium repair surgery. By their narrow definition, open heart surgery requires the use of a heart-lung machine and an incision into the heart so that at least one of its chambers is exposed.

The heart keeps oxygen flowing through the body. The right side of the heart receives deoxygenated blood from the body and pumps it to the lungs where it picks up oxygen. The oxygen-carrying blood returns to the left side of the heart where it is pumped out into the body. The problem with operating on the heart is that to keep the body supplied with oxygen, the heart must keep pumping, but operating on a beating heart is difficult to impossible. If the heart is to be stopped for surgery, a device must substitute for its function or the patient will die within a few minutes.

In 1931, John Heysham Gibbon (1903–1973) watched one of his patients at Boston Massachusetts General Hospital die slowly when a large clot blocked the artery that carries blood to the lungs causing the oxygen level of the blood to drop until there was no longer enough oxygen to support life. This experience made him wonder if there was a way to bypass the clot by removing some of the deoxygenated blood so that it could be refreshed with oxygen and returned to the body.

After several years of experimentation, Gibbon and his coworker and future wife, Mary Hokinson, built a device to oxygenate blood. The machine worked by exposing a thin layer of blood on a rotating cylinder to oxygen, collecting the oxygenated blood in a glass container that kept it at body temperature through submersion in a water bath, and then using a pump to return the blood to the body. Gibbons experimented on cats. On May 10, 1935, he used this apparatus to keep a cat alive for 39 minutes. Eventually, he was able to keep a cat alive for 2 hours and 51 minutes.

World War II interrupted Gibbon's work, but in 1945 after the war ended, he resumed his research in Philadelphia at Jefferson Medical College (now the Sidney Kimmel Medical College at Thomas Jefferson University). Through a medical student at Jefferson, he met Thomas Watson (1874–1956), the chairman of the board of IBM. Watson provided extensive funding for Gibbon's work. In addition to financing Gibbon's research, Watson had many of the specialized parts Gibbon needed for his machine made by machinists and technicians at IBM.

The first person Gibbon operated on was a 15-year-old girl. The machine worked perfectly, but the girl died because an incomplete diagnosis led the surgeons to look for her heart defect in the wrong place. The first successful open heart surgery using Gibbon's heart-lung machine occurred on May 6, 1953, when he repaired a heart defect in an 18-year-old woman. Unfortunately, his next two patients died, and Gibbons halted operations using the machine.

Meanwhile in Minneapolis, Minnesota, at the Variety Club Heart Hospital, Clarence Walton Lillehei (1918–1999) took another approach to open heart surgery. He used what he called cross-circulation. In this method, another human was used to oxygenate the blood of the patient being operated on.

On March 26, 1954, Lillehei successfully repaired what would otherwise have been a fatal heart defect in eight-month-old Gregory Glidden using cross-circulation. Gregory's father served as the oxygenating agent. Baby and father were anesthetized and connected by tubes and a pump. Deoxygenized blood flowed out of a vein in the baby and into a vein in the father. Oxygenized blood was pumped out of the father and back into the baby. This connection lasted for 19 minutes while the surgeon repaired the defect. The father survived the operation without any complications. The baby survived and was alert and eating, but seven days after the operation he developed pneumonia, and four days later he died.

Within 14 months, Lillehei had used cross-circulation to operate on 28 patients of which 62% survived. Many of these patients lived at least another 6 years. Regrettably, Lillehei became increasingly eccentric and his behavior in the operating room was considered dangerous. Because of this, his medical career was terminated in 1973.

Even though cross-circulation made open heart surgery possible, there was still a need for a machine to oxygenate blood and replace the volunteer human oxygenator. Richard A. DeWall (1926–2016), a colleague of Lillehei, developed a portable bubble oxygenator in 1955 that could be commercially manufactured. From then on, the heart-lung machine, technically known as cardiopulmonary bypass machine, was modified and improved.

As of 2018, about one million open heart surgeries were performed worldwide each year. Some of these are off-pump or beating heart surgeries where a heart-lung machine is not required. Others use minimally invasive surgical techniques but still require the use of a heart-lung machine. Complete heart or lung transplants are possible only because a machine can take over the role of a beating heart.

FURTHER INFORMATION

Meisner, Hans. "Milestones in Surgery: 60 Years of Open Heart Surgery." *Thoracic and Cardiovascular Surgeon* 62, no. 8 (2014): 645–650. https://www.thieme-connect.com/products/ejournals/abstract/10.1055/s-0034-1384802

PBS. "Partners of the Heart." *The American Experience*, February 10, 2003. https://www.pbs.org/wgbh/americanexperience/films/partners

Provident Foundation. "History." 2014. https://provfound.org/index.php/history

World History Documentaries. "Dr. Daniel Hale Williams First Black Heart Surgeon in America." YouTube, March 17, 2020. https://www.youtube.com/watch?v=JfUopc-pC_I

24. First Breast Augmentation Surgery, 1893

The first breast augmentation surgery was done on November 24, 1893, by Vincenz Czerny (1842–1916), a German surgeon living in what is now the Czech Republic. Czerny was the son of a pharmacist. He was expected to follow his father into that profession, but instead he enrolled in the University of Prague to study zoology and transferred to the University of Vienna for his medical education. At that time, Vienna had one of the most well-respected medical faculties in the world.

In Vienna, Czerny was fortunate to work with forward-thinking physicians who went against conventional beliefs. They taught the antiseptic procedures developed by Ignaz Semmelweis (1818–1865) years before these procedures were recognized by Joseph Lister (see entry 18). The emphasis on handwashing and sterilizing equipment helped Czerny develop a reputation as a successful gynecological and cancer surgeon.

In 1893, a 41-year-old opera singer came to Czerny because she had a lump that caused swelling and pain in her left breast. Czerny diagnosed a large tumor and recommended its removal. The woman was reluctant to have the operation because it would leave her with one breast significantly smaller than the other. She believed this would hurt her stage career. Nevertheless, when the pain and swelling continued, she sought a second opinion and then agreed to have Czerny remove the tumor.

While examining the woman before surgery, Czerny noticed a large noncancerous fatty tumor (a lipoma) on her lower back. At the time of the operation, he removed the breast tumor and also the tumor on her back. Then, while she was still under anesthesia, he used tissue from the fatty back tumor to replace tissue he had removed from her breast.

The woman spent 26 days in the hospital while both wounds healed without infection. At six months, she reported that her left breast was still tender. At one year, she said her left breast was pain-free. It felt slightly firmer and smaller than her right breast, but she was satisfied with the result. Czerny published a description of the tumor removal and replacement operation in 1895.

Was the surgery Czerny performed breast augmentation or breast reconstruction? Breast augmentation refers to a procedure done for cosmetic reasons. In this case, the woman was concerned about her appearance. She believed her career could be jeopardized by breasts of uneven size. Breast reconstruction refers to a procedure done after surgery for breast cancer. A tumor was the original reason for visiting Czerny and for having the operation. Depending on how the situation is interpreted, the procedure Czerny performed could be considered as either the first breast augmentation surgery or the first breast reconstruction surgery.

Czerny maintained a lifelong interest in treating cancer patients. At around age 50, he visited cancer hospitals in Buffalo, New York, and Moscow, Russia. On his return, he helped to found the Institute for Cancer Research in Heidelberg, Germany, which was open from 1906 to 2020. Czerny died of leukemia at age 74 after radiation treatment failed to cure his cancer.

HISTORICAL IMPACT AND CONTEXT

Vincenz Czerny did not originate the idea of using fat to replace surgically removed tissue. Gustav Albert Neuber (1850–1932), a German surgeon now mainly remembered for his insistence on extreme cleanliness and a separate sterile operating room, had reported success in transplanting a small amount of fat to fill a defect in the bone around the eye of a young woman. However, Neuber had cautioned that failures had occurred in other transplant surgeries when large amounts of fat were used.

Czerny was familiar with Neuber's work. When he saw the fatty tumor on his patient's back after she had expressed concern about her appearance, he was inspired to try the transplantation. That the fat came from the woman herself and Czerny's strict attention to sterility likely contributed to his success. It appears that he did not routinely perform similar operations.

In 1899, American surgeon William Stewart Halsted (1852–1922) performed the first radical mastectomy. This approach to breast cancer is now rarely used. It involves removal of the entire breast, the lymph nodes under the arm, and the chest wall muscles under the breast. Halstead believed that breast reconstruction interfered with control of the cancer and might allow it to redevelop undetected. He publicly opposed breast reconstruction, and his opinions were held in such high esteem that few American surgeons attempted the procedure.

Around the same time Halsted was speaking out against breast reconstruction after cancer surgery, women were seeking ways to enlarge their breasts for cosmetic reasons. During the first half of the twentieth century, surgeons tried implanting various materials to increase breast size. These included glass or wooden balls, ivory, ground rubber, ox cartilage, wool, and sponges made of various materials. None of these worked, and many caused significant harm that required surgical intervention and breast removal.

Between 1945 and 1950, surgeon Milton I. Berson (1897–1973) and Jacques Maliniac (1889–1976), a founder of the American Society of Plastic Surgeons, developed a technique to surgically rotate some of a woman's chest wall muscle into her breasts to enlarge them. This worked because no foreign material was involved, but it was a marginal solution. At about the same time, other surgeons tried using some newly developed synthetic materials as breast fillers. An estimated 50,000 women received direct breast injections of liquid silicone with disastrous results. Many developed a condition called silicone granuloma in which the breasts hardened and had to be completely removed.

The first real breakthrough in breast augmentation occurred when two plastic surgeons, Thomas Cronin (1906–1993) and Frank Gerow (1929–1993), worked with Dow Corning Corporation to produce a confined silicone implant. It consisted of a firm silicone bag filled with softer silicone gel. The implant was not considered a medical device and consequently did not need approval from the Food and Drug Administration (FDA). The surgeons tested the product on a dog and then in 1962 on a woman named Timmie Jean Lindsey in Houston, Texas. She was reportedly delighted with her newly enlarged breasts. After testing the implants on another 12 patients, they applied for a patent, which was granted in 1964. A successful marketing campaign began immediately.

The initial implants were found to be too firm to feel natural, so in 1972 Dow Corning introduced a new implant with a thinner bag and less viscous silicone filling. These more-flexible implants sometimes leaked, resulting in health problems and lawsuits that contributed to Dow Corning's bankruptcy in 1995.

More companies entered the booming implant business. In 1985, one manufactured a silicone bag that encased a bag of saline. The idea was that even if saline leaked, it was harmless and would not cause medical complications. In 1988, the FDA declared that breast implants were medical devices and needed proof of safety and FDA approval before they could be marketed.

Increased complaints of complications caused a temporary ban on silicone implants in 1992 that was soon lifted, but only for implants used in breast reconstruction and not for implants used in breast augmentation. A series of studies about the health effects of breast implants began in 1997 that led to the lifting of the ban on implants for augmentation in 2006 with FDA guidance that breast implants do carry medical risks and that women who have them should periodically undergo an MRI to check for leakage or damage.

The shapes of the implants have changed to meet changing tastes in body image. Textured implants were developed to produce a more natural feel, but in 2019, the FDA's General and Plastic Surgery Devices Advisory Panel insisted that one brand of textured implants be removed from the market because of its association with health complications.

In 2018, 1.86 million breast augmentations were performed worldwide. In 2021, breast augmentation was the second most common body aesthetic procedure performed in the United States, topped only by liposuction. According to the Aesthetic Society, that year 365,000 women had breast augmentation and another 101,657 women had breast reconstruction surgery. In the same year, 148,000 women had breast implants removed and replaced and 71,000 had their implants removed and not replaced.

The cost of breast reconstruction may be covered by insurance after cancer surgery, but the cost of a cosmetic procedure such as breast augmentation rarely is. Implants are not a one-time expense. They are designed to last at least 10 years and may last as long as 20. The chance of rupture increases by 1% each year, which is why yearly MRIs are recommended. Should rupture occur, there is a significant chance of health complications and additional expense to have the ruptured implant removed.

FURTHER INFORMATION

Goldwyn, Robert M. "Vincenz Czerny and the Beginnings of Breast Reconstruction." *Plastic and Reconstructive Surgery* 61, no. 5 (May 1978): 673–681. https://www .arslanianplasticsurgery.com/wp-content/uploads/sites/63/2017/04/beginnings-of -breast-reconstruction.pdf

Majumder, Sanjib. "Medicine in a Nutshell: The History of Breast Implants." YouTube, September 10, 2016. https://www.youtube.com/watch?v=Epa7_s5r5WE&t=53s

U.S. Food and Drug Administration. "7 Key Things You Should Know about Breast Implants." YouTube, August 20, 2020. https://www.youtube.com/watch?v=FxX4F5VByD

Zheng, Margaret. "The Development of Silicone Breast Implants for Use in Breast Augmentation Surgeries in the United States." The Embryo Project Encyclopedia, January 13, 2020. https://embryo.asu.edu/pages/development-silicone-breast-implants -use-breast-augmentation-surgeries-united-states

25. First Medical X-ray, 1895

Many medical advances are built on years of work by many people. Not so with X-rays. They were discovered through a single incident in 1895 by Wilhelm Conrad Roentgen (also spelled Röntgen, 1845–1923), a German physicist. Roentgen accidentally stumbled on X-rays while studying a different phenomenon in his laboratory on November 8, 1895. Fortunately, he recognized the significance of his discovery, and within days of his announcement, X-rays had changed the practice of medicine.

Wilhelm Roentgen was born in Lennep, Germany, but moved to the Netherlands at age three. He attended boarding school and then enrolled in Utrecht Technical School in Utrecht, Netherlands. At this school, he was falsely accused of drawing a caricature of one of his teachers and was expelled. Because of his expulsion, he could take university classes in the Netherlands but could not enroll in a degree program. Instead, he passed the entrance exam to the Federal Polytechnic Institute in Zurich, Switzerland, and in 1869 earned a PhD from the University of Zurich. After working at several universities, he became Chair of Physics at the University of Würzburg. While studying what happened when electricity passed through gas in a low-pressure tube, he discovered X-rays.

At the time Roentgen discovered X-rays, other researchers such as Nicola Tesla (1846–1943), William Crookes (1832–1919), and Philipp von Lenard (1862–1947) were experimenting with cathode rays. A cathode ray is a beam of electrons generated by passing an electric current between two electrodes sealed

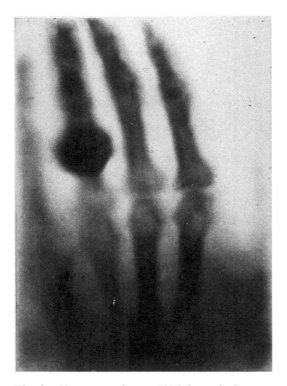

The first X-ray ever taken in 1895 shows the bones in the hand of Mrs. William Roentgen. On one finger is her wedding ring. Within weeks, X-rays were being used for medical diagnoses. (Courtesy of the National Library of Medicine)

in a near-vacuum tube. Electrons from the cathode (the negative electrode) cause the glass of the tube to glow green. Lenard had developed a glass tube covered in cardboard with a thin aluminum window at one end that allowed cathode rays to exit the tube. In a dark room, the energy from the cathode ray would create a slight glow on a screen treated with calcium sulfide when it was placed a few inches away from the tube.

On November 8, 1895, Roentgen was experimenting with cathode rays using a tube like the one Lenard had developed. Roentgen darkened the room and turned on the electric current. He expected to see a glow from energy exiting through the aluminum window in the tube. To his surprise, he saw a glow on a chemically treated screen several feet away, and the glow was coming from the side of the tube. How could this be? Cardboard was known to block cathode rays. The glow had to be produced by something else. Roentgen called these X-rays, the X standing for "unknown."

During several weeks of intense secretive work, Roentgen experimented by putting objects of different densities in front of the X-ray. He discovered that the degree of penetration depended on the density of the object in the ray's path. He also concluded that the glow was caused by high-energy ray. It was later confirmed that X-rays are electromagnetic waves that have a wavelength about 10,000 times shorter than the wavelength of visible light.

One day while holding a lead disc in front of the ray, Roentgen saw on the screen not just the disc but the bones of his hand holding the disc. This inspired Roentgen, an amateur photographer, to wonder if the mystery ray would develop a photographic plate. It did. On November 22, 1895, Roentgen asked his wife,

Bertha, to hold her hand over a photographic plate for 15 minutes while he exposed it to X-rays. (Unlike today's X-rays that can be made in seconds, early X-rays required a long exposure.) The resulting X-ray photo clearly showed the bones in Bertha's hand as well as the ring she was wearing.

Roentgen wrote up his discovery and on December 28 sent it to the Physical Medical Society of Würzburg. Almost immediately he received reprints of his article, and on January 1, 1896, he mailed 90 copies of the article along with some of his X-ray photographs to physicists across Europe. On January 5, the largest newspaper in Vienna, Austria, published some of the X-ray photographs on its front page. This caused a sensation. Physicists and doctors scrambled to buy tubes to make their own X-rays, and the price per tube rose to $20 (about $560 in 2022). In a matter of days, the worlds of medicine and physics had changed forever.

HISTORICAL CONTEXT AND IMPACT

Before the discovery of X-rays, it was impossible to see inside the human body without making a surgical incision. X-rays provided a whole new way to diagnose patients. Doctors started using them immediately. The first recorded diagnosis by X-ray occurred on January 6, 1896, just a few days after Roentgen released his findings to the world. A woman at Queen's Hospital in Birmingham, England, complained of pain in her hand. Fortunately, the hospital had the equipment required to make an X-ray, and this X-ray enabled a surgeon to remove a needle that was causing the pain. In North America, the first surgery to be guided by an X-ray occurred at Montreal General Hospital on February 7, 1896, when surgeons removed a bullet that had become embedded between the two bones of the lower leg of a man named Tolson Cunning. Cunning later used the X-ray as evidence to sue the shooter. Soon, others were using X-ray evidence in malpractice suits against doctors.

X-rays were also used in other ways. Within five months of their discovery, the British Army ordered two X-ray machines to help locate bullets in their soldiers fighting in Egypt, and in World War I, Marie Curie (1867–1934) developed a motor vehicle equipped with an X-ray machine that could be driven to battlegrounds. The X-ray machine derived the needed electrical current from the car's engine. X-rays were also used to diagnose lung diseases such as tuberculosis. This was possible because the rays penetrated diseased tissue and normal tissue to different degrees. Soon, doctors discovered that X-rays also had some therapeutic effects. Exposure to X-rays could cure certain skin diseases, and they were highly effective in relieving pain caused by cancer.

With early X-ray machines, the patient had to be exposed to X-rays for long periods—14 hours in one case to locate a bullet in the brain—in order to get usable pictures. As more people were diagnosed using X-rays, it became apparent that there were drawbacks to their use. Long exposure without protective equipment caused burns, tissue swelling, and hair loss. Although it was not apparent at the time, extended exposure to X-rays also caused cancer, and a substantial number of early doctors, researchers, and patients eventually developed cancer and died. By 1910, X-ray practitioners had become aware of the dangers of X-ray exposure. They very gradually began using protective equipment, but it was not until a more efficient X-ray tube was developed that exposure times could be decreased.

William Coolidge (1873–1975), working at the General Electric Research Laboratory in Schenectady, New York, developed a new type of X-ray tube in 1913 that solved many of the limitations of earlier machines. At General Electric, Coolidge had been experimenting with ways to draw the metal tungsten into fine filaments to use in light bulbs. Coolidge developed a low-cost way of doing this and then set about experimenting with other applications for the metal.

One problem with the X-ray tubes of Roentgen's time was that the tubes could not be a complete vacuum. The collision of gas molecules with the cathode was a necessary step in creating X-rays. This made the tubes highly variable, especially as the platinum cathode degraded with time. In what became known as the Coolidge tube, the cathode was made of tungsten instead of platinum. Tungsten has the highest melting point of any metal. When it is heated with an electric current to a very high temperature, X-rays can be generated in a complete vacuum. This improved the reliability of the tube and solved several other performance issues of older X-ray tubes.

One limitation the Coolidge tube could not overcome was that X-rays produced only two-dimensional pictures. This problem was solved in 1971 by Godfrey Hounsfield (1919–2004), an English electrical engineer, and Allan MacLeod Cormack (1924–1998), a South African–born American physicist. They developed X-ray computed tomography, usually referred to as a CT scan (see entry 47). In a CT scan, multiple X-rays are taken from different angles and converted into electrical signals. The signals are then compiled by a computer to produce a three-dimensional image. Hounsfield and Cormack shared the Nobel Prize in Physiology or Medicine in 1979. Thus, in less than 100 years, the rays that were accidentally discovered by Roentgen gave the world tools to see inside the body without surgery.

FURTHER INFORMATION

American Physical Society. "This Month in Physics History: November 8, 1895: Roentgen's Discovery of X-Rays." APS News 10, no. 10 (November 2001): 2. https://www.aps.org/publications/apsnews/200111/history.cfm

Doctor Kiloze. "History of X-Rays." YouTube, December 7, 2013. https://www.youtube.com/watch?v=fHUzVqoDnts

Kathy Loves Physics & History. "Physics of How Wilhelm Roentgen Discovered X-Rays." YouTube, September 20, 2018. https://www.youtube.com/watch?v=1PwxDEdl2iI

Queijo, Jon. "I'm Looking Through You: The Discovery of X-Rays." In Breakthrough!, 91–114. Upper Saddle River, NJ: Pearson Education, 2010.

26. First Electric Hearing Aid, 1898

Although people had been using passive devices to funnel sound toward the hard of hearing since the 1600s, it took the development of the telephone to create the first electric hearing aid. The first telephones built by Alexander Graham Bell (1847–1922) worked poorly and did not amplify sound. This changed in 1885 when Thomas Edison (1847–1931) developed a carbon transmitter for use in the telephone. The carbon transmitter would amplify weak sound vibrations when an electric current was passed through it.

Miller Reese Hutchinson (1876–1944) recognized that carbon transmitters could be adapted to devices to amplify sound for people with hearing loss. Hutchinson was born in Montrose, Alabama. As a child, he worked in machine shops and electrical repair shops. His mechanical skills and keen interest in electricity resulted in his receiving his first patent at age 16. He eventually earned a doctorate in electrical engineering at Alabama Polytechnic Institute (now Auburn University).

Hutchinson had a friend, Lyman Gould (1872–1937), who had lost his hearing after contracting scarlet fever as a young child. While studying electrical engineering, Hutchinson took some medical school classes to learn about the anatomy of the ear, and in 1898, after four years of research and experimentation, he created the first electric hearing aid for Gould. He named his device the Akouphone.

If you showed Hutchinson's Akouphone to people today, it is unlikely that they would identify it as a hearing aid. The device consisted of a separate microphone, an amplifier, headphones, and a large battery whose usage life was extremely short. The Akouphone was so bulky and heavy that it had to sit on a table and was barely portable. Still, it did give Gould some ability to hear.

After the relative success of the Akouphone, Hutchinson and a man named James H. Wilson founded the Akouphone Company. This was the first company specifically started to make hearing aids. The first commercial Akouphone was expensive—$400 in 1900, or about $12,700 in 2022—and very few were sold. Hutchinson completely remodeled the device to make it more portable and brought the cost down to about $60 ($1,950 in 2022). A third remodeling effort in 1902 resulted in the Acousticon. This model was much more portable. It consisted of a round carbon microphone, a battery, and a headband with earphones. Although it had no volume control, the Acousticon was remarkably successful. Hutchinson even went to Europe to fit Queen Alexandra of Denmark, the wife of King Edward VII of England, with the device. Alexandra claimed the Acousticon restored 90% of her hearing, and Hutchinson was richly rewarded.

Hutchinson and Wilson sold their company in 1905, and under its new ownership, the company continued to introduce improved hearing aids until it closed in 1984. Meanwhile, Hutchinson went on to develop the Klaxon Horn that makes a loud ah-HOO-gah sound. General Motors adopted the horn for use in their automobiles in 1912. Later, Hutchinson worked closely with Thomas

Edison on improving battery storage capacity and on issues of sound transmission and recording. In 1918, the association with Edison ended when—according to legend—Hutchinson became too intimate with a member of Edison's family. However, for another 25 years, Hutchinson continued inventing and investing in real estate and died a wealthy man.

HISTORICAL CONTEXT AND IMPACT

The earliest written mention of a hearing device in Europe can be found in a book written in 1634 by Jean Leurechon (1591–1670), a French Jesuit priest and mathematician. In this book, Leurechon described an ear trumpet. Early ear trumpets were funnel-shaped devices made of animal horn, wood, or metal. The wide, flared end gathered sound that was conveyed to the narrow end that the individual held to the ear. There was no mechanical amplification of the sound. Ear trumpets were made to order for each individual until 1800. In that year, F. C. Rein & Sons of London became the first commercial manufacturer of ear trumpets and other nonelectric hearing aids such as speaking tubes and acoustic urns. They continued to make ear trumpets until 1963.

During the 1800s, people with hearing loss were often assumed to have other disabilities, so it was common to camouflage acoustic hearing aids. One famous example is a chair built by F. C. Rein in 1819 for King John VI of Portugal. The chair had arms that ended in lion heads with open mouths. These contained concealed resonators that gathered sound, funneled it through a hollow tube that rimmed the back of the chair, and fed the sound into a flexible tube the king could subtly insert into his ear. To improve sound collection, the king insisted that anyone speaking to him must kneel so as to be near the open chair arms. Acoustic urns, another type of hearing aid, were often placed on tables and disguised with flowers or fruit. There were even acoustic walking sticks. None of these solutions were very satisfactory, and the hard of hearing were often shunned, isolated, and considered mentally deficient.

Although Miller Hutchinson invented the first electric hearing aid, the first commercially successful device using electronic amplification was manufactured by Siemens Corporation in 1913. Early electric models were larger than a cigar box, awkward to use, and did not work particularly well. They had no volume control and no way to screen out environmental noise. Batteries used to provide the needed electricity were heavy and had short lifespans. But hearing loss was common, and there was a steady market for these hearing aids despite their limitations.

The development of the vacuum tube, which could substantially amplify sound waves, started a generation of improvement in hearing aids. The first patent for a vacuum tube hearing aid was granted in 1920 to Earl C. Hanson (1892–1979), a California electronics wizard and inventor. He developed what he called the Vactuphone. It was manufactured commercially beginning in 1921 by a joint effort between Western Electric and Globe Phone Manufacturing Company. The device weighed seven pounds and cost $135 ($2,235 in 2022).

As vacuum tubes became smaller, so did hearing aids. The first commercially available wearable vacuum tube hearing aid appeared in England in 1936 and in the United States in 1937. In these early wearable vacuum tube hearing aids, the battery pack was worn around the neck or hidden in a purse and the microphone was hand-held, making the wearer look like today's television reporter doing an on-site interview.

The next major improvement in hearing aids came with the development of the transistor at Bell Laboratories in 1948. Transistors were tiny compared to vacuum tubes. They also used less power, which extended battery life. The use of transistors seemed like a great leap forward in hearing technology. Unfortunately, early transistors were sensitive to body heat and dampness, and the first generation of transistor hearing aids consistently failed after a few weeks. The problem was solved in 1954 when Texas Instruments produced a more resilient silicon transistor. This allowed the development of hearing aids that were incorporated into eyeglass frames and models that could be worn discretely behind the ear.

During the 1960s, Bell Telephone Laboratories created a way to digitally process speech and audio sound on computers, but the mainframe computers needed to do this were much larger and much slower than today's computers. What hearing aid manufacturers needed was a low-power, lightweight way of processing audio signals in real time. The first fully digital experimental hearing aid was created in 1984 at City University of New York, but it was impractical for actual use. A commercial digital hearing aid made by Nicolet Corporation in 1987 was such a complete failure that the company went out of business. Finally in 1996, after years of advances in high-speed digital array processors and microcomputers, a practical, commercially successful digital hearing aid called the Senso was developed by Widex.

In 2020, the National Institute on Deafness and Other Communication Disorders estimated that 15% of Americans over age 18 have some trouble hearing. Today these people have the option of using practically invisible digital hearing aids that can be worn in the ear canal and are customizable and programmable. These hearing aids have directional microphones, can screen out most distracting noise, amplify sound within the range where the individual's hearing loss is greatest, and self-adjust to different sound environments. The most modern of them can be used with smartphones and Bluetooth-enabled devices so that sound can be streamed directly into the ear. Because of advances in hearing aid technology, hearing deficits are no longer a reason to isolate individuals from full participation in society.

FURTHER INFORMATION

Bernard Becker Medical Library. "Deafness in Disguise: Concealed Hearing Devices of the 19th and 20th Centuries." Washington University School of Medicine, May 14, 2012. https://beckerexhibits.wustl.edu/did/index.htm

British Pathé. "Hearing Aid Museum." YouTube, April 13, 2014. https://www.youtube.com/watch?v=RiCU3OrKlQ8

27. First Recognized Human Virus, 1900

In 1881, Cuban physician Carlos Juan Finlay (1833–1915) presented a paper to the Havana Academy of Sciences suggesting that yellow fever, a common and often fatal disease, was transmitted by the bite of an infected mosquito. This idea was so extraordinarily different from the common theories about disease at the time that no one took Finlay's idea seriously.

Carlos Finlay was born in Cuba, the son of a Scottish physician and a French woman. He studied in both France and England, but the University of Havana refused to recognize his European education, so he enrolled in Jefferson Medical College in Philadelphia. He graduated in 1855 and spent two years in Europe acquiring specialized medical training before returning to Cuba in 1857.

Yellow fever had regularly devastated Cuba and other parts of the Caribbean and South America since the mid-1600s. There was, and still is, no cure for the disease. In 1879, the United States National Health Board Yellow Fever Commission visited Cuba to study the disease. Finlay was invited to join the Commission. The Commission concluded incorrectly that yellow fever was transmitted by air. Finlay was unsure about their conclusion and continued to study yellow fever for the next 20 years. He noted that an increase in cases regularly corresponded to the wet season in Havana when the number of mosquitoes increased. He wondered if mosquitoes, not air, were carrying the disease from person to person.

Finlay could not find any experimental animals that developed yellow fever, so he tested his mosquito theory on 102 human volunteers. From these experiments, he became convinced that mosquitoes transmit the disease. Even after his human experiments, Finlay's theory was ridiculed. Doctors were reluctant to accept the new idea that yellow fever was not passed directly from person to person but needed an insect to act as an intermediary. Resistance to the mosquito theory was especially strong because Finlay could not isolate the organism that caused the disease or grow it in a culture.

Proof that Finlay's mosquito theory was correct came in 1900 when the United States Army Yellow Fever Board, under the direction of Walter Reed (1851–1902), visited Cuba. Finlay was involved in their visit and, along with Walter Reed, convinced one board member, James Carroll (1854–1907), to volunteer to be bitten by an infected mosquito. Carroll developed a mild case of yellow fever and recovered. Another board member, Jesse Lazear (1866–1900), was accidentally bitten and died. This convinced Reed that Finley's theory was correct. Careful filtering of fluid from infected mosquitoes also indicated that the disease was caused by a particle smaller than a bacterium. Using this information, the army started a massive mosquito control program. Cases of yellow fever in Havana dropped dramatically.

HISTORICAL CONTEXT AND IMPACT

Finlay's credibility problem arose partly because yellow fever is caused by a virus. Viruses, unlike bacteria, are too small to be seen under a light microscope, and they were barely recognized in Finlay's time. In 1892, Dmitry Ivanovsky (1864–1920), a Russian plant scientist, became the first person to isolate a virus. While studying the source of a tobacco plant disease common in southern Russia, he isolated the tobacco mosaic virus.

Ivanovsky put sap from infected tobacco leaves through a recently developed filter that was fine enough to strain out bacteria. When he injected the bacteria-free filtrate into healthy tobacco plants, the plants became infected. Ivanovsky, like Finlay, did not have a name for what he had found. He could not see the organism that he had isolated, but he understood that it was not a traditional bacterium or fungus.

The yellow fever virus causes chills, headaches, nausea, and fever. After a period of relatively mild illness, some people recover, but up to one-quarter of those infected appear to improve and then relapse and develop severe symptoms, including internal bleeding. Many of the people who relapse die. Until Finlay linked the disease to mosquitoes, outbreaks of yellow fever were common in port cities in the United States. In 1793, yellow fever swept through Philadelphia killing 10% of the population and sickening many more. During the Spanish-American War (1898), five times more American soldiers died from yellow fever in Cuba than died in combat. The effect of the disease on combat readiness was a major reason why the army got involved in tracking down the cause of the disease and initiating a mosquito elimination program. Today yellow fever is rare in the United States. The last major outbreak on American soil occurred in New Orleans in 1905.

Perhaps the greatest impact of the discovery that yellow fever is caused by a virus transmitted by mosquitoes was on the building of the Panama Canal. In 1881, the French under the leadership of Ferdinand de Lesseps (1808–1894) began to build a canal across the Isthmus of Panama. The canal was of worldwide importance. It would provide prestige and income for France and would save ships of all nations thousands of miles and days of travel. Lesseps had successfully led the team that built the Suez Canal, and he expected the Panama Canal to be easier. What he had not taken into consideration was the toll that yellow fever, and to a lesser extent malaria, would take on workers.

Between 1881, when work began, and 1898, when the French finally abandoned work on the canal, 22,000 workers died from yellow fever and malaria. Tens of thousands more became sick; over 85% of workers required hospitalization. Retaining laborers became difficult. Many healthy workers simply ran away out of fear of the disease.

In 1904, the United States took over building the canal. William C. Gorgas (1854–1920), an army physician who had been part of the Army Yellow Fever

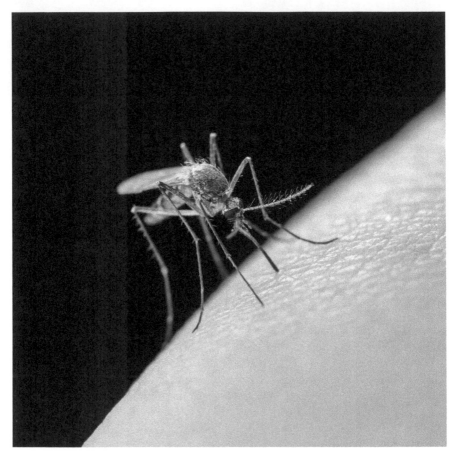

Carlos Finlay spent more than 20 years proving that yellow fever, the first recognized
human virus, was transmitted by the bite of an infected mosquito and not by "bad air."
(Konstantin Nechaev/Dreamstime.com)

Board that had proved Finlay's theory about mosquito transmission was correct,
was put in charge of eradicating mosquitoes in the Canal Zone.

Gorgas used 4,000 men working full-time for a year on mosquito abatement.
He had houses sprayed with insecticide and fitted with screens. He eliminated
pools of standing water. Hundreds of tons of insecticide were used in fumiga-
tion. By the end of one year, there were only a handful of yellow fever cases in
the Canal Zone, and after 1906, there were none. The canal was completed and
opened on August 15, 1914. The total number of deaths from all illnesses and
accidents under the Americans was 5,609.

Despite progress in understanding human viruses, their transmission, and the
diseases they caused, their structure remained elusive until 1931 when Max Knoll

(1867–1969) and Ernst Ruska (1903–1988) invented the electron microscope in Berlin. This microscope used a beam of electrons that could be focused in a way that let the user see objects smaller than those seen with visible light. The first commercial electron microscope became available in 1939. For the first time, viruses could be visualized.

Seeing viruses was one thing. Preventing them from causing illness and death was another. Some of the world's worst diseases—including smallpox, HIV/AIDS, Ebola virus disease, and COVID-19—are caused by viruses. Antibiotics, the mainstay of treatment against bacterial diseases, are ineffective against viruses. Vaccines, including the one against yellow fever, can stimulate the immune system to attack a specific virus, but vaccines work only if they are given a week or two before someone becomes infected and remain effective only if the virus does not rapidly mutate. With a better understanding of viruses, the search was on for drugs that could destroy them.

William Prusoff (1920–2011) described the first drug with antiviral properties in 1959. The drug interfered with DNA synthesis in cancer cells, but Prusoff recognized that it also showed antiviral activity against some herpes viruses. Unfortunately, this drug and other early antiviral drugs had unacceptable side effects.

Until the early 1970s, antiviral drugs were discovered by trial and error. This changed when Gertrude Elion (1918–1999) and George Hitchens (1905–1998) used an understanding of the virus's genetic material to develop drugs specifically designed to cause a virus to make faulty copies of its genes. These faulty genes would prevent the virus from reproducing. Their breakthrough drug was acyclovir. Hitchings, Elion, and Sir James Black shared the 1988 Nobel Prize in Physiology or Medicine for fundamentally changing the thinking about how to design new drugs. Elion went on to work on other antiviral drugs including AZT, the first successful drug to treat HIV/AIDS (see entry 54). Antiviral drugs are still limited in terms of the infections they can treat, and more than 100 years after recognizing the yellow fever virus, there is still no drug to cure the disease.

FURTHER INFORMATION

Adams, Patrick. "Meet the Woman Who Gave the World Antiviral Drugs." *National Geographic*, August 31, 2020. https://www.nationalgeographic.com/science/article/gertrude-elion-antivirals-coronavirus-remdesivir

Faerstein, Eduardo, and Warren Winkelstein Jr. "Carlos Juan Finlay: Rejected, Respected, and Right." *Epidemiology* 1, no. 21 (January 2010): 158. https://journals.lww.com/epidem/Fulltext/2010/01000/Carlos_Juan_Finlay__Rejected,_Respected,_and_Right.28

PBS. "The Great Fever." *The American Experience*, October 30, 2006. https://www.pbs.org/wgbh/americanexperience/films/fever/#transcript

Science Universe. "History and Discovery of Viruses." YouTube, December 11, 2020. https://www.youtube.com/watch?v=3U30utSOBRI

28. First American Food and Drug Purity Laws, 1906

Before the American Civil War (1861–1865), most people raised their own food or bought it from local farmers. After the war, the population of cities increased. City residents did not have land to raise their own food. At the same time, improved rail transportation made it possible to move food over great distances, but for this food to reach cities unspoiled, much of it needed to be preserved. These factors gave rise to an increasingly important but unregulated food processing industry. Meanwhile, drugs remained equally unregulated. They were often ineffective or contained harmful ingredients such as heroin, mercury, and arsenic.

These conditions gave rise to the Pure Food and Drug Act of 1906 and the Meat Inspection Act. Both Acts were signed on the same day and were the first comprehensive federal food and drug regulatory laws in the United States. But these Acts would not have been passed without the work of Harvey Washington Wiley (1844–1930), chief chemist of the Bureau of Chemistry, and journalist and author Upton Sinclair (1878–1968).

Harvey Wiley grew up on an Indiana farm. He attended college for one year but left to join the Union Army. After the war, he finished college, earned a medical degree, taught at Purdue University, and was appointed the state chemist of Indiana. As the state chemist, he analyzed commercial sugars and syrups to determine if they were adulterated. He also became concerned about "embalmed milk" and "embalmed beef," products to which formaldehyde was added to keep these foods "fresher" longer. Embalmed beef later caused a scandal when it was sent to troops in Cuba during the Spanish-American War (1898). The meat was so heavily adulterated and poorly preserved that it was toxic, and soldiers died after eating it.

In 1882, after being passed over for the position of president of Purdue University—allegedly because he was too young, too friendly, had odd religious beliefs, and was unmarried—Wiley took a job as chief chemist at the United States Department of Agriculture. There he continued to study food additives and adulterated products. To do this, he recruited in 1902 what became known as the Poison Squad. This was a group of volunteer men who for $5 per month ($150 in 2022) agreed to eat specially prepared food containing various doses of chemical food preservatives. They tested borax that was used to prolong the life of rotting beef, as was salicylic acid and sodium sulfite. The men became so sick during the sodium sulfite trial that the experiment was stopped.

Wiley used the results of the Poison Squad experiments to bring the problem of food adulteration to the attention of the public and to push for a law to control the use of these substances. He was aided in his crusade by Upton Sinclair (1878–1968), a muckraking journalist who graphically exposed the disgustingly unsanitary conditions in the meatpacking industry in his book *The Jungle* (1906), and by Fanny Farmer (1857–1915), author of a famed cookbook who promoted the use of pure, unadulterated food by home cooks. The result was the Pure Food and Drug Act, which was signed into law by President Theodore Roosevelt on June 30, 1906. The passage of the Act was highly contested. The *New York Times* said the debate was one of the wildest of the congressional session. The Federal Meat Inspection Act, vigorously opposed by the meatpacking industry, was passed and signed the same day.

HISTORICAL CONTEXT AND IMPACT

In 1875, Massachusetts passed a law intended to assure the quality of food sold in the state. It provided for fines, imprisonment, and public shaming by standing in the stocks for anyone who knowingly sold "unwholesome provisions," but the law applied only to Massachusetts. There was no movement toward a national food and drug purity law until 1879, when Peter Collier, head of the Bureau of Chemistry (which later evolved into the Food and Drug Administration), investigated food adulteration and pushed for a national food and drug purity law. An appropriate law came before Congress in 1880 and was defeated. It took another 26 years for the Pure Food and Drug Act and the Federal Meat Inspection Act to be passed.

The Meat Inspection Act required the U.S. Department of Agriculture to inspect livestock before slaughter and to inspect food products derived from livestock after slaughter. It prohibited the slaughter of sick animals, required that animals be killed under sanitary conditions, and outlawed common adulterants in canned and fresh meat. The law was substantially overhauled and strengthened in 1967.

The Pure Food and Drug Act barred interstate commerce in misbranded and adulterated foods, drinks, and drugs. At the time, many foods and drugs contained harmful ingredients such as morphine, cocaine, alcohol, sulfites, and chemical dyes. The law required accurate labeling of all food and drugs. It did not require premarketing testing or approval of drugs. The strength of the law lay in a provision for confiscation and destruction of foods or drugs found by inspectors from the Board of Food and Drug Inspection to be in violation of the law. Manufacturers were encouraged to comply because of the threat of lost profits if their products were destroyed.

Many lawsuits challenging the interpretation of the 1906 Act followed. In 1912, the U.S. Supreme Court ruled in *Johnson v. United States* that false and misleading therapeutic drug claims were not "misbranding." The Congress promptly passed the Sherley Amendment to overcome this ruling and eliminate false claims.

The first major overhaul of the 1906 Act came with the Food, Drug, and Cosmetic (FDC) Act of 1938. The FDC Act extended regulations to cosmetics and therapeutic devices and required that drugs be shown to be safe before marketing. They did not have to prove that they were effective. In the Wheeler-Lea Act the same year, the task of overseeing appropriate advertising of FDA-regulated products was given to the Federal Trade Commission. In 1943, in the case of *United States v. Dotterweich*, the Supreme Court ruled that company executives could be held responsible for violating the 1938 Act irrespective of whether the violations were intentional or even known to the executives.

Although various amendments were passed to the 1938 Act, the next major overhaul occurred with the 1962 Kefauver-Harris Amendment, known as the Drug Efficacy Amendment. This amendment required for the first time that drugs had to be proven effective as well as safe before being marketed. The driving motivation behind this amendment was the finding that thalidomide, a new drug used as a tranquilizer and to prevent nausea in pregnant women, caused severe birth defects in their babies. The pill was available in 46 countries. Although the drug was kept off the American market by the FDA, its effects in Europe drove the need for stronger drug safety regulations.

Since then, various amendments have been added to keep up with advances in science and medical technology. The Medical Device Amendment in 1976 was passed to assure safety and effectiveness of medical devices and diagnostic tests, some of which require premarketing approval. The 1994 Dietary Supplement Health and Education Act stated that dietary supplements such as herbs, vitamins, and minerals should be treated as foods, not drugs. This freed these products from premarket approval for safety and efficacy and led to a boom in natural and alternative medicines. Both the 1997 and the 2011 Food and Drug Modernization Acts streamlined the review of drugs and medical products, changed advertising regulations, and provided that the same standards be applied to imported foods as apply to domestically produced foods.

Current criticisms of the food and drug regulatory process in the United States are that regulation is confusingly spread across too many federal organizations and that some of these organizations cannot keep up with the rapid changes that are occurring, especially in biologic therapeutics. For example, the regulation of genetically modified organisms (GMOs) has come before the Congress, but as of 2023, there was no federal legislation specifically addressing GMOs. GMO products are confusingly regulated by three different agencies: the Department of Agriculture's Animal and Plant Health Inspection Service, the FDA, and the Environmental Protection Agency. Biologics—that is vaccines, blood products, gene therapy products, tissue transplantation, and new immunotherapy treatments for cancer—are regulated by the Center for Biologics Evaluation and Research, a part of the FDA. Rapid advancements, especially in the area of immunotherapy, have stressed this organization to keep up with new product regulations and approvals, especially for well-publicized advances that the public anticipates as beneficial. Despite these drawbacks, foods and drugs sold in the United States remain some of the safest in the world.

FURTHER INFORMATION

Bennis, Camille, Alayna Brightbill, and Lyndsey Ramberg. "Pure Food and Drug Act 1906." YouTube, February 9, 2019. https://www.youtube.com/watch?v=xDsGcvwPHN4&t=152s

Blum, Deborah. *The Poison Squad: One Chemist's Single-Minded Crusade for Food Safety at the Turn of the Twentieth Century.* New York: Penguin, 2018.

Constitutional Rights Foundation. "Upton Sinclair's *The Jungle*: Muckraking the Meat-Packing Industry." *Bill of Rights in Action* 24, no. 1 (Fall 2008). https://www.crf-usa.org /bill-of-rights-in-action/bria-24-1-b-upton-sinclairs-the-jungle-muckraking-the-meat -packing-industry.html

Meadows, Michelle. "Promoting Safe and Effective Drugs for 100 Years." *FDA Consumer Magazine*, January–February 2006. https://www.fda.gov/about-fda/histories-product -regulation/promoting-safe-effective-drugs-100-years

29. First Human Laparoscopy, 1910

The first documented human laparoscopy was performed by physician Hans Christian Jacobaeus (1879–1937) in Stockholm, Sweden. Little has been written in English about Jacobaeus's early life and training. In 1910, he published a paper reporting on his initial experiments with human laparoscopy. Laparoscopy is a surgical procedure in which a tube (cannula) containing a tool is inserted through a very small incision into the abdomen for diagnostic or therapeutic purposes. It is similar to an endoscopy, except that in an endoscopy the tube and tools are inserted through a natural body opening such as the mouth, rectum, or vagina and no surgical incision is needed.

The device Jacobaeus used was quite different from the modern laparoscope. It consisted of a cannula that contained a sharp three-pointed blade called a trocar. Jacobaeus's only sources of illumination were candles and natural light. Using this primitive laparoscope, Jacobaeus operated on 17 patients who had extensive fluid buildup (ascites) in the abdomen. The trocar was used to penetrate the skin and abdominal wall, and the excess fluid was removed through the cannula. Jacobaeus reported that he withdrew 8–10 liters of fluid (2–2.5 gallons) from most patients, but in one patient, he removed 23 liters (about 6 gallons). Despite the relief his patients must have felt, his report was not well received by the medical community and few other physicians showed any interest in using the procedure.

One exception was German physician Georg Kelling (1886–1945). After Jacobaeus published his findings, Kelling claimed to have performed human laparoscopies earlier than Jacobaeus. It turned out, Kelling had performed documented laparoscopies on dogs in 1901, but he had not published any results on humans before Jacobaeus—and the scientist who publishes first is generally credited as the inventor of a new procedure.

Jacobaeus saw great potential in the use of laparoscopy and ignored the criticism of the medical community. He next reported on 97 laparoscopies. Out of the 97, only 8 patients did not have large accumulations of abdominal fluid. Without the presence of this fluid to distend the abdomen, Jacobaeus judged the risk of puncturing an internal organ too high to justify the procedure. If he accidentally nicked the intestine, leaked material would cause internal infection, which in the pre-antibiotic era usually resulted in death.

In the patients without excess abdominal fluid, Jacobaeus experimented with injecting air into the abdomen. Injecting air was a technique Kelling had pioneered. The air, sealed in the body by a flap he added to the cannula, helped to separate the internal organs and reduce the chance of accidentally cutting into one of them. The experiments with air led to the first thoracoscopy, a laparoscopic intervention in the chest cavity. Jacobaeus also successfully used injected air to help release abnormal attachments of the lung to the chest wall in patients with tuberculosis.

From 1925 until his death in 1937, Jacobaeus was a member of the Nobel Prize Committee for Physiology or Medicine. Based on documents released by the Nobel Prize organization, he blocked António Egas Moniz (1874–1955), a Portuguese neurologist, three times (1928, 1932, 1937) from receiving a Noble Prize for his method of visualizing brain blood vessels. It is unclear what Jacobaeus's objections were. Egas Moniz's technique was revolutionary and remained in use until CT scans were developed. Ironically, Egas Moniz won the Nobel Prize in 1949 for developing the prefrontal lobotomy as a surgical treatment for mental illness. This treatment is unacceptable today.

HISTORICAL CONTEXT AND IMPACT

Jacobaeus's experiments with human laparoscopy evolved from an invention in 1806 by German physician Philipp Bozzini (1773–1809). Bozzini wedged open the urethra with a speculum and set up a candle and angled mirrors to focus the light and reflect the interior of the urethra. This allowed him for the first time to observe the interior of the body without cutting into it. He examined the vagina in the same way. Bozzini called his device a *Lichtleiter* or light conductor. This was an awkward contraption, but it took until 1835 for another physician, Frenchman Antonin Jean Desormeaux (1815–1894), to improve the design by replacing the candle with an alcohol lamp and the speculum with a hollow tube. He called his device an endoscope.

Adolph Kussmaul (1822–1902), another German physician, wanted to study the upper part of the gastrointestinal tract. His endoscope consisted of a gasoline lamp, angled mirrors, and an 18.5-inch-long (47 cm), 5.1-inch-diameter (13 cm) rigid tube. He enlisted a professional sword swallower to swallow the tube and was able to see the esophagus and upper part of the stomach. Regrettably, his patients did not have the same ability to swallow the device as the sword swallower, so this research was abandoned. Jacobaeus's primitive laparoscope evolved from these early endoscopes.

Laparoscopies were introduced in the United States in 1919, but few physicians were interested in using them. In the 1920s, Walter C. Alverez (1884–1978), a San Francisco physician, successfully used carbon dioxide to inflate the abdomen. This gas is absorbed by the body and exhaled through the lungs much more rapidly than air or oxygen. It soon became the standard gas for abdominal inflation. An improvement in safety occurred in 1938 when Hungarian Janos Veres (1903–1979) produced a safer needle. Once the needle had penetrated the body cavity, a spring was released that covered its tip and protected the surrounding tissue from accidental puncture.

Despite these advances, diagnostic laparoscopy was slow to be accepted. A 1966 survey of American internists and surgeons found that only 10% had ever performed a diagnostic laparoscopy and fewer than 1% had done more than 50. The general thinking among surgeons was that there was no reason to do a diagnostic laparoscopy when with open (large incision) surgery the condition could be diagnosed and corrected during the same surgical session.

In the next decade, gynecologists were the first specialists to accept the use of laparoscopy as both a diagnostic and therapeutic procedure. Laparoscopic surgery was effectively banned in Germany from 1956 to 1961. Nevertheless, during the 1970s, German gynecologist Kurt Semm (1927–2003) reported that he had performed 3,300 laparoscopic surgeries. He had used the procedure to remove ovarian cysts, remove ovaries, treat tubal pregnancies, and perform female sterilizations (tubal ligations).

Many traditional conservative gynecologists criticized Semm and refused to believe his report, but Semm converted his more broadminded colleagues and convinced them of the value of the technique. One study found that in 1971, only 1% of sterilized women had had tubal ligation done through laparoscopic surgery. By 1975, 60% of tubal ligations were done laparoscopically. The laparoscopic procedure allowed faster recovery, left behind a much smaller scar, and was no more risky than open surgery.

FURTHER INFORMATION

Franciscan Health. "History of Minimally Invasive Surgery." YouTube, June 4, 2018. https://www.youtube.com/watch?v=ReFZHY3Rggs

Litynski, Grzegorz. "Laparoscopy—The Early Attempts: Spotlighting Georg Kelling and Hans Christian Jacobaeus." *Journal of the Society of Laparoscopic & Robotic Surgeons* 1, no. 2 (January–March 1997): 83–85. https://www.ncbi.nlm.nih.gov/pmc/articles /PMC3015224

Nakayama, Don K. "The Minimally Invasive Operations that Transformed Surgery." American College of Surgeons Poster Competition, 2017. https://www.facs.org/-/media /files/archives/shg-poster/2017/10_minimally_invasive.ashx

30. First Vitamin Identified, 1912

Vitamins are substances necessary in tiny quantities for human health. The body cannot make these micronutrients; they must be acquired from food. An absence of a particular vitamin results in a specific disease. For example, in 1747 James Lind (1716–1794), a Scottish naval doctor, performed an experiment that showed the disease scurvy could be prevented by the addition of citrus fruit to sailors' diets (see entry 8). Lind, however, had no idea that the protective substance in citrus fruit was vitamin C.

In the early 1900s, researchers experimented with inducing different diseases in animals by feeding them diets that withheld certain foods. They could produce a disease and reverse it with specific foods, but once again they had no way of identifying the essential disease-preventing component in that food. Umetaro Suzuki (1874–1943), a Japanese researcher at Tokyo Imperial University, was the first person to isolate one of these essential substances, a water-soluble micronutrient found in rice bran that would prevent beriberi. Beriberi is a disease that causes muscle stiffness, strange eye movements, pain in the hands and feet, and eventually paralysis. Suzuki named the new substance aberic acid. It was later identified as vitamin B1 (also called thiamine). Suzuki published his research in Japanese in 1910. When his work was translated into German, at the time the common language of scientific research in Europe, the translator failed to include the information that the isolated substance was a *new* nutrient, although this fact was in the original Japanese manuscript. As a result, Suzuki's research generated little attention.

Shortly after Suzuki published, Polish biochemist Casimir (also spelled Kazimierz) Funk (1884–1967) began working on beriberi. Funk, a Polish Jew, was born in Warsaw, the son of a well-known dermatologist. He attended the University of Bern in Switzerland and graduated with a PhD in 1904. After this, he worked in labs in Paris and Berlin before moving to the Lister Institute in London where he was assigned to work on beriberi. Here he found that feeding rice bran—the brown outer husk of rice that is removed during polishing—to pigeons could reverse experimentally induced beriberi. He then had the insight that people who ate white rice were more likely to develop beriberi than people who ate brown rice with the bran intact. He concluded that there was an essential substance in rice bran that prevented the disease.

Funk managed to extract what he thought was the essential nutrient in bran. What he actually extracted was vitamin B3 (niacin), which is essential in preventing pellagra, not beriberi. In 1912, Funk published a book, *Die Vitamines*, in which he proposed the existence of other essential micronutrients in food. He called these vital substances vitamines, "vita" meaning life and "amine" for the nitrogen group he believed they contained. The "e" in "amine" was dropped in 1920 when it was shown that not all essential compounds contained amines. From then on, vitamines were known as vitamins.

Funk bounced back and forth between the United States and Europe several times, holding various industry and university positions related to research on vitamins, hormones, and cancer. He went on to identify vitamins B1, C, and D. Because he was the first European to isolate and name these essential compounds and to write a widely read book about them, Casimir Funk, rather than Umetaro Suzuki, is considered the father of vitamin therapy.

HISTORICAL CONTEXT AND IMPACT

Ancient Chinese, Greek, and Roman cultures were all familiar with vitamin deficiency disorders and had ideas about how certain foods could heal and balance the body. Their theories were based on observation and philosophies about the nature of food and the nature of illness. Sometimes, they even worked. For example, ancient Egyptians recognized that night blindness, a condition caused by vitamin A deficiency, could be alleviated by eating liver. Today we recognize liver as a good source of vitamin A.

It took until the late 1800s for researchers to show concrete connections between diet and specific diseases. Beriberi was common among Japanese sailors whose diet consisted mainly of white rice. General Kanehiro Takaki (1849–1915) believed wrongly that beriberi was caused by protein deficiency and ordered that Japanese sailors be provided with a more varied diet and more protein. The increased variety mostly eliminated beriberi but did nothing to uncover its cause. Dutch physiologist Christiaan Eijkman (1858–1930), who worked in the Dutch East Indies (present-day Indonesia), discovered that chickens fed only white rice developed beriberi and that the disease could be reversed by feeding them brown rice or rice bran. He misguidedly believed that a poison in white rice was neutralized by something in the bran that was removed when the rice was polished. Eijkman's contribution to the science of nutrition was belatedly recognized in 1929 when he and Frederick Hopkins (1861–1947) were awarded the Nobel Prize in Physiology or Medicine for vitamin research. At the time, some in the nutrition community felt Casimir Funk had been unjustly overlooked.

With recognition of the essential nature of micronutrients in food, the race to identify new vitamins took off. From 1913 to 1922, six new vitamins were isolated and their food sources identified. One of the most important was vitamin A. Elmer McCollum (1879–1967) at the University of Wisconsin and his often-ignored assistant, Marguerite Davis, beat out by three weeks Lafayette Mendel (1872–1935) and Thomas Osborne (1859–1929) who were working in Connecticut by first reporting the discovery of vitamin A. McCollum later discovered B vitamins and vitamin D, earning him the nickname Dr. Vitamin in the popular press. McCollum is also remembered in scientific circles for starting the first rat colony for the study of nutrition. Nevertheless, he left the University of Wisconsin under a cloud, accused of claiming credit for another researcher's work, stealing research notebooks, and sabotaging research by releasing all the rats from their cages when he left the university.

Without a doubt, vitamins saved lives and prevented disability in people with limited diets. Pellagra (vitamin B3 deficiency), once epidemic across the American South where poor people existed on a corn-based diet, was practically eliminated. Rickets (vitamin D deficiency) also practically disappeared. Along with these benefits came the commercialization of vitamins. In 1920, Parke-Davis and Company produced the first multivitamin capsule, claiming that it contained concentrated extracts of vitamins A, B, and C.

Some vitamin products had value while others were worthless, but Americans bought into the promise that vitamins would provide better health and vitality in a big way. Although the government warned about fraudulent vitamin promises, it also recognized that many Americans ate a poor diet. To counter this, vitamin D–fortified milk was introduced in the early 1930s, and flour fortified with vitamins B1 and B3 arrived at the end of that decade. In 1941, the government released the first recommended dietary allowances (RDAs), which included the daily requirements for six vitamins and two minerals.

In the 1950s, vitamins were intentionally packaged in apothecary bottles to look like drugs in order to enhance their image of effectiveness and reliability. Today there are 13 recognized essential vitamins, and many major pharmaceutical companies such as Abbott Laboratories, GlaxoSmithKline, and Pfizer have highly profitable vitamin divisions.

Vitamins in the United States are regulated under the 1994 Dietary Supplement Health and Education Act in the same way that food is regulated. Like food manufacturers, makers of vitamins do not have to prove that a supplement is safe or effective before it can be sold to the public, but they are limited in the specific health claims they can make. Although manufacturers may claim that vitamins improve health, they cannot claim that vitamins cure, treat, or prevent any specific disease.

As of 2019, the American vitamin and supplement market was estimated at $35 billion annually, with the annual worldwide market at $128 billion. One in three American children take a daily vitamin, and 77% of all Americans use some sort of dietary supplement, the most common being vitamins and minerals. For all the good vitamins can do, they are needed only in tiny quantities; larger doses are not better. As vitamin usage has increased, so have warnings from the healthcare community about the dangers of mega-doses. Despite the fact that most Americans fulfill their vitamin needs through their diet, Americans still gulp down daily vitamins with the idea that they will promote better health.

FURTHER INFORMATION

Carpenter, Kenneth J. "A Short History of Nutritional Science: Part 3 (1912–1944)." *Journal of Nutrition* 133, no. 10 (October 2003): 3023–3032. https://academic.oup.com /jn/article/133/10/3023/4687555

Nostalgic Medicine. "A Brief History of Vitamins." YouTube, April 5, 2021. https://www .youtube.com/watch?v=rJN_eVpWlek

Rosenfeld, Louis. "Vitamine—Vitamin. The Early Years of Discovery." *Clinical Chemistry* 43, no. 4 (April 1, 1997): 680–685. https://academic.oup.com/clinchem/article/43/4/680/5640821

31. First Country to Legalize Abortion, 1920

Abortion—the intentional termination of a pregnancy—has been performed in every civilization from the earliest of times. In almost all cultures, authorities either ignored the practice, failed to substantially enforce rules against it, or drove it underground by criminalizing the procedure and punishing both the woman who had the abortion and the person who performed it. In 1920, Russia became the first country to formally legalize abortion-on-demand.

Nicholas II was emperor of All Russia in 1917. He was a highly conservative autocrat completely out of touch with the people he ruled. He was supported by the equally conservative Russian Orthodox Church. In this environment, abortion under any circumstance was considered a serious punishable crime. This did not stop the practice, as many underground or "back room" abortions were performed by midwives and older women who had some nursing training.

In 1917, a populist uprising forced Nicholas II to vacate the throne. This marked the end of 300 years of conservative rule by the House of Romanov. After a period of bloody civil war among various political factions, the Bolsheviks formed a government headed by Vladimir Ilyich Ulyanov (1870–1942), better known as Lenin. In 1920, under Lenin's leadership, the Russian Soviet Federative Socialist Republic, which would become the Soviet Union in 1922, legalized abortion-on-demand at any point in a woman's pregnancy.

The rationale for legalizing abortion was to create a more egalitarian society. The Bolsheviks believed that abortion-on-demand would give women control over their reproductive rights and free them from the need to marry for financial support by recruiting them into the workforce where they could earn an income. In addition, state-performed abortions would improve women's health by being safer and more sanitary than backroom abortions. The December 1920 Decree on Abortion stated that:

- Abortions were to be performed free of charge in Soviet hospitals.
- Only a doctor could perform the operation.
- Any midwife or nurse performing an abortion would lose the right to practice and be subject to a court trial.
- Any doctor who charged for a private practice abortion would face a court trial.

Abortion-on-demand was meant to be a temporary measure to improve women's health by improving their economic and social conditions, but, because almost no effective birth control was available, abortion came to be used as a routine method of birth control often to the detriment of women's health. Within a few years, hospitals were so overwhelmed by women having abortions that the government set up dedicated abortion clinics to free up hospital beds for ill patients.

Around the same time that hospitals and clinics were overflowing with abortion seekers, the government was encouraging women to have more children to offset a population decline caused by the loss of life in World War I and the civil war that followed the fall of the Tsar. Gradually, government restrictions were imposed on abortions. These changes began to limit abortions to the first three months of pregnancy unless there was a need for a later abortion. "Need," however, was loosely defined and included economic and social conditions. In addition, women who were poor or who already had a large family were given priority at abortion clinics, forcing some women to wait beyond the three-month on-demand period. These changes simply drove women back to using illegal abortion providers.

Finally, in 1936 during the reign of Joseph Stalin (1878–1953), abortions were declared illegal. This ended the first experiment with legal, on-demand abortion. However, in 1955, Nikita Khrushchev (1874–1971) made abortion legal again while still discouraging the procedure by emphasizing its dangers. Nevertheless, the difficulty in obtaining effective birth control caused the abortion rate to remain high in the Soviet Union. Even into the late 1980s, 70% of pregnancies in Russia were terminated by abortion.

HISTORICAL CONTEXT AND IMPACT

Writings from almost all ancient cultures show that abortive agents were familiar and frequently used. The most common method of abortion was for the woman to consume large amounts of specific herbs early in pregnancy. This often caused a miscarriage. Unfortunately, the quantity of herbs the woman needed to ingest was often toxic, and many women died. Aborting an older fetus by introducing objects into the uterus was even more dangerous. Women often bled to death or died of infection.

The practice of aborting pregnancies was so common in ancient Greece that the original Hippocratic Oath, which at the time was not binding on physicians, contained the promise not to give a woman a drug or physical object that would cause her to abort her pregnancy. Not everyone in ancient Greece agreed with this. Plato (428–347 BCE) and Aristotle (384–322 BCE) both believed it was acceptable to use abortion for population control. Roman law considered abortion a property crime against the husband who had been deceived by being deprived of the fetus, which was considered his property. Roman women who had abortions could be exiled or put to death. Only in ancient Persia was the fetus considered a person and its life protected by law.

For several hundred years, countries passed laws criminalizing or severely limiting abortion. These laws were enforced with varying rigor depending on the social attitudes of the time. In 1803, the United Kingdom made abortion before quickening (feeling the fetus kick) a felony and abortion after quickening a death penalty offense. Canada codified its laws in 1869 to make all abortions illegal. This remained the case until 1969 when some abortions were selectively permitted. In the United States, the American Medical Association strongly condemned abortion in 1859, making an exception for cases where it was to save the life of the mother. By 1879, every state had made abortion illegal. Despite the threat of serious punishment, desperate American women continued to have abortions illicitly. Many died of complications.

In the 1930s, some countries began modifying their strict antiabortion laws. Mexico legalized abortion in cases of rape in 1931. In 1935, Iceland legalized some abortions based on hardship. Sweden followed in 1938. Both Japan and Germany relaxed their abortion laws during World War II to promote their political theories of eugenics.

In the United States, abortion was regulated by state law. In the mid-1950s, it was illegal in every state. On May 21, 1959, the American Law Institute released influential recommendations for modifying American abortion laws. They recommended that abortions be legal to perform in hospitals in cases of rape or incest, in case of physical abnormality of the fetus, and for the physical or mental health of the mother. In 1966, Mississippi became the first state to permit abortion in the case of rape. The next year, Colorado became the first state to reform its laws based on the guidance issued by the American Law Institute. Within two years, 10 other states had followed Colorado's lead. Then in 1970, Hawaii law changed to allow abortion-on-request in early pregnancy, New York permitted the procedure up to the 24th week of pregnancy, and Washington state passed a referendum— with 56.49% of the vote—that permitted early-pregnancy abortions.

Although states were independently changing their abortion laws, national change came in 1973 when the U.S. Supreme Court ruled in *Roe v. Wade* that individual state bans on first-trimester abortions were unconstitutional, that states could regulate second-trimester abortions, and that they could prohibit third-trimester abortions unless the woman's physician ruled that a late abortion was necessary for her physical or mental health.

Change was rapid. In 1965, when abortion was still nationally criminalized, there were a reported 794 abortions. In 1974, the year after *Roe v. Wade*, there were 898,570. The number continued to climb, and from 1979 through 1993, there were on average 1.5 million abortions performed in the United States each year. After 1993, the numbers began dropping. In 2018, only 327,691 abortions were reported.

Abortion has always been politically, socially, and religiously controversial, so many new laws and lawsuits followed the Supreme Court decision. Laws affecting the procedure passed in state legislatures generally had restrictions to the procedure that became more stringent in response to social and political pressures. These laws served to narrow and limit the scope of the Supreme Court's ruling. They were usually challenged in lower courts. Most often, the laws imposed barriers such as forbidding public funding of abortions and reducing the time following conception during which a woman could legally abort the fetus. Some also required a combination of a waiting period, pro-life counseling, ultrasound visualization of the fetus, and/or invasive physical examinations before an abortion was performed.

On June 24, 2022, the U.S. Supreme Court decided the *Dobbs v. Jackson Women's Health Organization* case. The case challenged the state of Mississippi's ban on abortions 15 more weeks after conception. The Court ruled that the U.S. Constitution does not confer a right to abortion. This decision overturned the 1973 decision of *Roe v Wade* and the 1992 decision of *Planned Parenthood v. Casey*.

As a result of *Dobbs v. Jackson Women's Health Organization*, regulation of abortion was returned to the individual states, many of implemented stringent laws prohibiting abortion after six weeks of pregnancy with few or no exemptions for pregnancies arising from rape or incest. Some states had what are called trigger laws that automatically went into effect immediately after the Court's decision. These laws varied from state to state but they generally banned and criminalized abortions, both for the women who had them and the healthcare workers who provided them. Some states also passed personhood laws that declared a fetus at any stage of development to be a preborn person with equal personhood rights. As of 2023, when this book was written, abortion remains legal in some states, is heavily regulated in some states, and is strictly illegal in other states. The effects of these changes and the legal implications they present are working their way through the courts.

FURTHER INFORMATION

CNN. "Abortion Wasn't Always Taboo in America." YouTube, April 3, 2017. https://www.youtube.com/watch?v=7E6ojeOLVdQ

"*Dobbs v. Jackson Women's Health Organization*." Center for Reproductive Rights, June 2022. https://reproductiverights.org/case/scotus-mississippi-abortion-ban

Regan, Leslie J. *When Abortion Was a Crime: Women, Medicine, and Law in the United States, 1867–1973*. Berkeley: University of California Press, 1997.

"Termination of Pregnancy." Museum of Contraception and Abortion, 2020. https://muvs.org/en/topics/termination-of-pregnancy

32. First Physical Therapy Organization, 1921

The American Women's Physical Therapeutic Association, the first professional physical therapy organization, was established in 1921 by Mary McMillan

(1880–1959). She is considered the Mother of Physical Therapy. McMillan was born in Boston, Massachusetts, the daughter of recent Scottish immigrants. Her mother died when she was still a young child. When her father remarried, she was sent to live with relatives in Liverpool, England.

McMillan began university studies in Liverpool in 1900. She soon became interested in therapeutic movement and massage. At that time, exercise and massage were taught as part of the physical education curriculum. Believing her education was inadequate for understanding therapeutic movement, McMillan moved to London to study anatomy, neurology, and psychology. In the early 1900s, a few orthopedic surgeons privately trained women with some physical education or nursing background on how to use specific exercises to help children disabled by polio. Physical therapy (also called physiotherapy) as the profession we recognize today did not exist.

Injuries suffered by soldiers in World War I (1914–1918) increased the demand for people trained in therapeutic movement and joint rehabilitation. McMillan volunteered her services to the British Army but was turned down for health reasons, so she returned to the United States and volunteered to treat soldiers at Walter Reed General Hospital, the U.S. Army hospital in Washington, D.C. She was the first person to be given the formal Army title of Reconstruction Aide. Before this, there was no standard term for a woman (all reconstruction aides were women) who helped restore functional mobility through exercise and manipulation. At Walter Reed, McMillan developed new protocols to treat war injuries and taught them to the nurses. During her time there, she also took a leave of absence to train 200 women as reconstruction aides at Reed College in Portland, Oregon.

By 1921, World War I was over and the reconstruction aide program at Walter Reed was ending. McMillan wanted to keep intact a group of women with reconstruction aide skills. She persuaded interested women to meet at Keens Chophouse in New York City on January 15, 1921, to form the American Women's Physical Therapeutic Association, the first professional organization of physical therapists. The next year, the organization changed its name to the American Physiotherapy Association. In 1923, the first two men were admitted, and in 1928 the organization established minimum qualifications for the profession, including graduation from a school of physical education or nursing supplemented by 1,200 hours of physical therapy instruction. Another name change in 1947 resulted in the organization being called the American Physical Therapy Association (APTA), which remains its name today.

Mary McMillan spent her life as an ambassador for physical therapy. In 1932, she went to China under the auspices of the Rockefeller Foundation to establish a physical therapy department at Union Medical College in Beijing. She was in the process of returning to the United States by way of the Philippines when Pearl Harbor was bombed on December 7, 1941. Instead of returning home, McMillan volunteered her services at the Army hospital in Manila. When Manila was occupied by Japanese troops, she was taken a prisoner of war. In the prisoner-of-war camp, she set up a makeshift space where she held a daily physical therapy clinic despite becoming weak with beriberi, a vitamin B1 deficiency caused by poor diet (see entry 30). On her repatriation to the United States, McMillan continued as an ambassador for physical therapy for another 10 years.

HISTORICAL CONTEXT AND IMPACT

Although physicians from ancient Greece and ancient India recognized that exercise, massage, and even hydrotherapy were good for health, there was no organized system of teaching or applying these techniques. Pehr Henrik Ling (1776–1839), a Swedish educator, was the first person to establish a formal program that evolved into physical therapy. Ling was university-educated and traveled extensively in Europe. He learned to fence in France and discovered that exercise helped his gout and joint pain. He was also strongly influenced by the book *Gymnastics for the Youth* by Johann Christoph Friedrich GutsMuths (1759–1839), a German educator who introduced gymnastics, which he called "culture for the body," into the school curriculum.

On his return to Sweden, Ling took the classes needed to become a medical doctor but did not go on to practice medicine. Instead, he used his knowledge of anatomy and physiology to develop a series of science-based gymnastic exercises and manipulations that he called medical gymnastics. In 1813, he established the Royal Central Gymnastics Institute, now called the Swedish School of Sport and Health Sciences. For this, Ling is considered the father of Swedish gymnastics.

Ling's gymnastics curriculum did not include massage, which is part of today's physical therapy training. Johann Georg Metzger (1838–1909), a Dutch physician, introduced massage in Europe around 1870. In 1884, four British nurses established the Society of Trained Masseuses in order to professionalize massage therapy and counteract its association with prostitution. In 1920, the organization became the Society of Massage and Remedial Gymnastics, combining two aspects of modern physical therapy. Men were then allowed to become members. In 1944, the organization became the Chartered Society of Physiotherapy. Today it represents more than 53,000 physical therapists in the United Kingdom.

In the United States, World War II (1939–1945) and the Korean War (1950–1953) increased the need for physical therapists to help rehabilitate injured soldiers. At this time, physical therapists were still considered technicians who carried out treatments prescribed by a doctor, and the profession was still female dominated. For example, the Women's Medical Special Corps founded in 1947 admitted only female physical therapists. Male therapists were not allowed to join until 1955 when the name was changed to the Army Medical Specialists Corps.

During the 1950s, the APTA, in which Mary McMillan was still active, began advocating to change the perception of physical therapists from technicians to medical practitioners. Physical therapy was moving from orthopedic surgeons' offices into hospitals, a move accelerated by the passage of the Hill-Burton Act in 1946 that provided funding for modernizing and constructing new hospitals and other healthcare facilities. In response, the APTA developed a seven-hour-long competency test for physical therapists. The organization also successfully pushed for universities and medical schools to provide physical therapy training both at the undergraduate and graduate level, and in time a bachelor's degree became necessary for entry into the field.

Changes to the Social Security Act in 1967 gave another boost to physical therapy in the United States by requiring Medicare to pay for at-home or out-patient physical therapy that was administered at the direction of a hospital-associated physician. Along with these changes came increased educational and licensing requirements encouraged by the APTA. The shift from viewing physical therapists as technicians who carried out doctors' orders to viewing them as healthcare providers in their own right had begun. By the late 1970s, after 20 years of advocacy by the APTA, physical therapists in some states were allowed to diagnose and prescribe treatment independent of a medical doctor, and patients could refer themselves directly to physical therapists.

In January 2016, the Commission on Accreditation in Physical Therapy Education made Doctor of Physical Therapy (DPT) the required degree for all of its accredited entry-level physical therapist education programs. In addition, the graduate had to pass a licensing examination. The DPT requirement was controversial. It increased the time and cost of becoming a physical therapist, but it was supported by the APTA as a way to put physical therapists on par with other independent healthcare professionals. A DPT degree generally requires a four-year bachelor's degree plus three years of physical therapy instruction. The annual median income of DPTs in 2021 was about $91,000. The APTA now recognizes physical therapy specialization in 10 areas, including sports, geriatrics, neurology, orthopedics, and oncology. Specialty certification requires additional training and examinations.

Physical therapy assistants are also recognized by the APTA. They are state-licensed individuals with two years of education and training who must work under the supervision of a DPT. Their median salary in 2021 was about $60,000. As of 2019, there were almost 313,000 licensed physical therapists and 128,000 physical therapy assistants in the United States. Women still dominate the field. About 74% of licensed physical therapists are female.

FURTHER INFORMATION

Elson, Mildred O. "The Legacy of Mary McMillan." *Journal of the American Physical Therapy Association* 44, no. 2 (December 1964): 1066–1072.

Farrell, Mary, and Marta Mobley. *Mary McMillan—The Mother of Physical Therapy.* Independently published, 2002.

Shaik, Abdul R., and Arakkal M. Shemjaz. "The Rise of Physical Therapy: A History in Footsteps." *Archives of Medicine and Health Sciences* 2, no. 2 (July–December 2014): 257–260. https://www.amhsjournal.org/article.asp?issn=2321-4848;year=2014;volume =2;issue=2;spage=257;epage=260;aulast=Shaik

33. First Isolation and Purification of Insulin, 1922

Insulin is a hormone made by beta cells in the pancreas. It is released into the bloodstream as part of a system to regulate blood glucose (blood sugar) levels. All cells need glucose for energy. Without insulin, glucose cannot enter cells, so glucose builds up to life-threatening levels in the blood. When the pancreas makes inadequate amounts of insulin, a person develops diabetes. People who make some, but not enough, insulin develop type 2 diabetes. Type 2 diabetes can usually be treated with diet, exercise, and medication. People who make no insulin develop type 1 diabetes. Before 1922, when animal insulin was isolated and purified for human use, a diagnosis of type 1 diabetes was a death sentence.

The story of the isolation and purification of insulin involves four Canadian men at the University of Toronto: Frederick Banting (1891–1941), a farm boy who alienated his family by becoming a doctor; John James Rickard Macleod (1876–1935), an influential professor of physiology; Charles Best (1899–1978), a university student Macleod assigned to work with Banting; and James Collip (1892–1965), an outstanding biochemist on leave from the University of Alberta.

Frederick Banting was a shy, awkward man who established an unsuccessful medical practice in London, Ontario, after serving as a surgeon in the Canadian Army during World War I. On October 31, 1920, Banting happened to read a journal article about the pancreas. This was not his area of expertise, but it sparked an idea that became an obsession. Most of the pancreas is made up of cells that secrete digestive enzymes into a duct that empties into the small intestine, but the pancreas also contains small islets of cells that secrete insulin directly into the bloodstream.

In 1920, the pancreas's function was a bit of a mystery. Earlier research hinted that it had something to do with diabetes symptoms. Banting had the idea that if the duct to the intestine was tied off and the digestive enzyme–producing cells died, he could figure out the function of the islet cells.

Through an acquaintance, Banting arranged a meeting with Macleod. He wanted the professor to give him laboratory space and support at the University of Toronto to explore his idea. The meeting between the shy, inarticulate farm-boy doctor and the upper-class, snobbish Macleod was a disaster. Macleod dismissed Banting, telling him to research similar work that had already been done on this problem.

Banting was obsessed with his idea. He eventually wrote a formal research proposal and Macleod agreed to give him laboratory space for a summer and a dozen dogs to experiment on. He also assigned Best to be Banting's assistant. Macleod then left to spend the summer in Scotland. The research did not go well. Tying off the pancreatic duct was tricky surgery and several dogs died of infection. Banting and Best eventually resorted to searching the streets of Toronto late at night to collect stray dogs to use in their experiments. Another problem was that Banting was a surgeon and Best a student. They had only rudimentary biochemistry skills that were inadequate for isolating the substance in the islet cells.

Failure followed failure, but finally they saw some success. A collie whose pancreas had been removed was kept alive for 19 days by injecting it with islet cell extract from experimental dogs. The collie died when no more extract was available. Based on this achievement, Macleod grudgingly agreed to allow Banting to continue his experiments. Banting, who was deeply in debt because he had been working without a salary, then sold his house, its contents, and his medical practice in order to continue his research.

By November 1921, Banting and Best had enough successes for Macleod to suggest that Banting present their preliminary findings to the university's physiology club. Macleod was supposed to introduce Banting but instead did the entire talk, referring to the research as "ours," as if he had worked on the project while he was actually vacationing in Scotland. Banting was furious that Macleod would take credit for work he and Best had done. This was the start of lifelong animosity between the two men. The same thing happened the next month at a conference at Yale University. Macleod took over, using "we" and "our" to describe research he had not actively participated in.

Meanwhile, James Collip had been awarded a Rockefeller Fellowship that allowed him to work and study at other laboratories anywhere in the world for one year. He chose the University of Toronto. His area of research was gland secretions, which overlapped with Banting's work. Macleod had set up a second laboratory in a different building where he and Collip were also working on purifying islet cell extract. What was supposed to be a collaboration became a competition.

Banting and Best finally developed an extract pure enough to test on a human. On January 11, 1922, they injected Leonard Thompson, a 14-year-old diabetic who was near death. The extract caused only a small decrease in Thompson's glucose level, so the hospital refused to allow another injection. A few days later, Collip, who was a better biochemist than Banting or Best, announced that he and Macleod had found a new way to purify the extract and that he was going to leave the research group and patent the process. When Banting heard this, he physically attacked Collip. The men were separated by Best who happened to be in the lab. When Collip's extract was given to Thompson, his blood glucose dropped substantially. With daily insulin shots, Thompson lived another 13 years and eventually died of pneumonia.

For the next four months, both groups experimented on animals with Collip's insulin. And then a major setback occurred. Collip had committed a serious scientific sin. He had failed to write down his purification process in his lab notebook and by some quirk suddenly forgot some vital part of the isolation process. When he told Banting about this, Banting once again attacked him. For a long time after that, the two men could not be left alone in the same room.

Finally, Collip recreated his purification process. In April 1922, a patent was obtained on the product and all four men signed the patent over to the

University of Toronto for one dollar. On May 2, 1922, Macleod presented the insulin research results at the Association of American Physicians conference. Banting and Best did not attend.

Banting and Macleod were jointly awarded the Nobel Prize in Physiology or Medicine in 1923. Neither man attended the award ceremony, refusing to be on stage with the other. Banting still believed that Macleod had taken credit for his and Best's work. He was so angered that Best had not been included in the prize that he split his prize money with Best, embarrassing Macleod into doing the same for Collip.

HISTORICAL CONTEXT AND IMPACT

The news that diabetes was no longer a death sentence created a sensation and was reported by newspapers worldwide. The disease had been around since ancient times. It had been described in early Egyptian writings. In ancient India, physicians had learned to diagnose diabetes by exposing ants to the patient's urine. If the ants were attracted to sugar in the urine, the patient was diabetic. Despite accurate diagnosis, there was no cure.

Before there was insulin, physician Frederick Allen (1876–1964) had pioneered an agonizing treatment—the starvation diet. Allen discovered that a diet of about 500 calories daily would keep glucose levels under control. Every bit of food had to be weighed and measured. The patient was always voraciously hungry. Some starved to death. Others went off the diet and died. Allen, unlike many charlatans of the time, never claimed that he could cure diabetes, only claiming that his diet might keep the patient alive until a cure could be found.

Elizabeth Hughes (1907–1981) was one of Allen's successes. She was diagnosed with diabetes at age 11 and was on the Allen diet for two years. By age 13, she was almost a skeleton—five feet (152 cm) tall and weighing 48 pounds (21 kg). Once she was able to receive daily insulin shots, she gained weight, finished high school, graduated from college, married, had a child, and lived a long life.

Eli Lilly Company in Indianapolis, Indiana, produced the first commercial insulin in 1923. Almost immediately, researchers began experimenting with the insulin molecule for ways to provide a variety of strengths, alter the speed with which insulin would take effect, and change how long it would remain active. Regardless of the variety of insulin, until 1978 all insulin was extracted from pork and beef pancreases that were shipped by the refrigerated boxcar load from slaughterhouses to pharmaceutical companies.

In 1978, David Goeddel (b. 1951) and his colleagues at Genentech created the first synthetic human insulin using recombinant DNA technology. This insulin, called Humulin, was approved for use in the United States in 1983. Today all insulin is laboratory-made. There are multiple strengths and types of insulin, from short-acting to long-acting. There are also new delivery methods, including insulin pen injections and insulin pumps. Modern technology also allows people with diabetes to continuously monitor their glucose levels by a combination of internal monitoring, a transmitting device, and a smartphone. Worldwide, the human insulin market in 2020 was worth $27.71 billion and was expected to continue to grow.

FURTHER INFORMATION

Bliss, Michael. *The Discovery of Insulin*. Chicago: University of Chicago Press, 2007.

Cooper, Thea, and Arthur Ainsberg. *Breakthrough: Elizabeth Hughes, the Discovery of Insulin, and the Making of a Medical Miracle*. New York: St. Martin's, 2010.

Davidson, Tish. *Vaccines: History, Science, and Issues*. Santa Barbara, CA: Greenwood, 2017.

Nature Video. "The Discovery of Insulin." YouTube, June 18, 2021. https://www.youtube.com/watch?v=Gk1D4VgM8jY&t=251s

Wu, Brian. "History of Diabetes: Past Treatments and New Discoveries." *Medical News Today*, April 29, 2019. https://www.medicalnewstoday.com/articles/317484

34. First Antibiotic, 1928

Alexander Fleming (1881–1955), a British bacteriologist, discovered penicillin, the first identified antibiotic, on September 28, 1928. Fleming was born on an isolated farm in Scotland, the seventh of eight children. He attended a small local school but at age 11 won a scholarship to a larger school. Two years later, he moved to London to live with his brother, a physician, and attend school there.

After graduating, Fleming worked in a shipping office for four years until an inheritance from an uncle allowed him to attend St. Mary's Hospital Medical School in London. His intention was to become a surgeon after he graduated in 1906. Nevertheless, when a job opened up in the research department at St. Mary's, Fleming changed his plans and took the job where he worked under prominent bacteriologist Sir Almroth Wright (1861–1947), inventor of an antityphoid vaccine.

During World War I (1914–1918), Fleming served in the Royal Army Medical Corps doing laboratory research. His war service stimulated an interest in the treatment of wounds. He noted that antiseptic chemicals worked well on soldiers' surface wounds but harmed deep wounds by destroying tissue.

Following the war, Fleming continued his research at St Mary's where in 1921 he discovered lysosome, a mild antiseptic agent found in tears, saliva, mucus, and breast milk. In 1927, he began an investigation of staphylococcal bacteria responsible for serious infections such as pneumonia, infections of the skin, bone, and the tissue surrounding the heart. As part of his research, he grew staphylococcus in shallow plates coated with agar, a gelatin-like substance on which

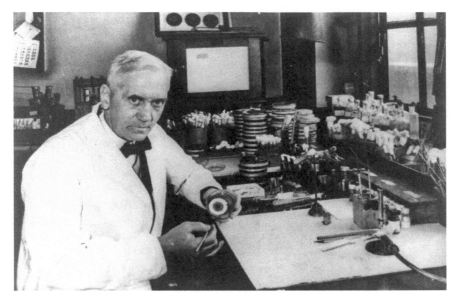

Alexander Fleming discovered penicillin in 1929, but it did not become a drug of interest until World War II when controlling infection in wounded soldiers became a priority.
(Library of Congress)

bacteria will grow. The plates were covered to prevent cross-contamination from
microorganisms in the air.

Fleming set up a new set of agar plates before he went on vacation. On his
return, he found that one had been left uncovered and was contaminated with
mold. Myth says that the mold spores came in through an open window, but it is
much more likely that they came through the ventilation system from the laboratory on the floor below where researchers were working with fungi.

Fleming's great contribution to medicine came about not through intentional experimentation and observation but because he was curious. Instead
of washing the contaminated plate, he examined it closely and noted that no
staphylococcal bacteria were growing near the mold, but farther away from the
mold, the bacteria had grown abundantly. Interested in why this happened,
he experimented and discovered that a liquid extract from the mold would
kill many types of bacteria. After several months of referring to the extract as
"mould juice," he named it penicillin after *Penicillium*, the genus to which the
mold belonged.

Fleming published the results of his discovery in 1929, but neither he
nor other researchers saw any immediate medical use for the new substance,
and his findings generated little interest. *Penicillium* mold was hard to grow
in the quantities needed to extract a usable amount of penicillin. In addition, although Fleming was a good bacteriologist, neither he nor his assistants

were biochemists. No one in his laboratory had the skills needed to isolate and purify the mold extract. Fleming continued to do a few experiments with penicillin but concentrated on other research. He was close to retirement when a team of scientists figured out how to convert Fleming's "mould juice" into a lifesaving drug.

HISTORICAL CONTEXT AND IMPACT

Many fungi and some plants have antibiotic properties, and ancient cultures including those in India, Egypt, Greece, and China commonly used plants and fungi to treat infections, with variable success. In seventeenth-century Europe, moldy bread was applied to wounds, but at that time there was no concept of microorganisms causing disease so there was no understanding of why this treatment sometimes worked. In fact, when Fleming discovered penicillin, the word "antibiotic" did not exist. It was coined around 1940 by Nobel Prize winner Selman Waksman (1888–1973), a soil bacteriologist at Rutgers University in New Jersey who discovered more than 20 antibiotics, including neomycin and streptomycin, both of which have widespread medical applications.

In 1938, Fleming's almost-10-year-old publication came to the attention of a group of scientists at Oxford University that included Australian Howard Florey (1898–1968), German Jewish immigrant Ernst Chain (1906–1979), and British biochemist Norman Heatley (1911–2004). World War II had just started. From past wars, it was well known that more soldiers died from infected injuries than were killed outright in battle. Finding a way to control infection was in the national interest.

By 1939, the Oxford team was involved in intense penicillin research, but there was a problem. It took 2,000 liters of mold culture fluid (what Fleming called mould juice) to extract enough penicillin to treat a single person. The laboratory grew mold in every container that it could find, including bathtubs, bedpans, and milk churns and hired women known as "penicillin girls" to tend to the mold.

Eventually, the team extracted enough penicillin to inject 10 mice with staphylococcus. Five were treated with penicillin and five were left untreated. All the treated mice lived. All the untreated mice quickly died. The experiment was an unqualified demonstration of the value of penicillin. Then in February 1941, a policeman, Albert Alexander, who had been scratched with a rose thorn while gardening and developed an infection that spread to his eyes, face, and lungs, became the first person to be given penicillin. He immediately improved but later died when the supply of the drug ran out before he was completely cured.

It was essential to find a way to mass-produce penicillin. Britain was at war and did not have the facilities to produce the amount needed. This led Florey to try to interest American pharmaceutical companies in the drug. At first he was unsuccessful, but eventually Florey connected with a U.S. Department of Agriculture laboratory in Peoria, Illinois. Here scientists found a different species of *Penicillium* that produced greater quantities of penicillin. Exposing this species to X-rays to cause mutations eventually produced a species that made 1,000 times more penicillin than Fleming's mold. The lab also found a more efficient way to grow the fungi, and mass production became possible. In 1943, American pharmaceutical companies began producing the drug under the direction of the War Production Board.

Initially almost all penicillin went to the troops, and its success was indisputable. The rate of death by infection dropped from 18% to 1%. Dr. Chester Keefer (1897–1972) of Boston was appointed "civilian penicillin czar." His role was to ration civilian use of the drug, which meant he had the unenviable job of turning down frantic pleas from the families of dying loved ones. By the end of World War II, American pharmaceutical companies were producing 650 billion units of penicillin a month, and rationing was no longer necessary.

In 1945, Fleming, Florey, and Chain were jointly awarded the Nobel Prize in Physiology or Medicine. Norman Heatley was excluded because Nobel Prize rules allow no more than three people to share the prize. Heatley's contribution was finally recognized by Oxford University in 1990 when they awarded him an honorary doctorate of medicine. He was the first individual with a nonmedical education to receive this honor in 800 years.

Antibiotics quickly became the most successful class of drugs on the planet. In 2021, the global market for antibiotics was estimated at $40.7 billion. They are prescribed liberally by doctors in developed countries and are widely used in veterinary medicine. Millions of pounds of antibiotics have been given to animals raised for food, both to prevent infection and to encourage growth.

Bacteria can mutate rapidly, and some will develop mutations that make them resistant to antibiotics. Widespread use of antibiotics has produced a rapidly increasing number of bacteria species that are resistant to multiple classes of antibiotics. By one estimate, 40% of some species of bacteria are highly resistant to *all* current antibiotics. The path to new drug development is long, slow, and expensive, and the search for effective new antibiotics cannot keep up with the rate at which bacterial resistance is developing. Only 15 new antibiotics were approved in the United States between 2000 and 2018, and most of them were variations of already known types of antibiotics.

Recent government programs and public pressure have encouraged doctors and veterinarians to prescribe fewer antibiotics and farmers to raise antibiotic-free animals, but antibiotic resistance continues to increase. In 2018, the U.S. Centers for Disease Control and Prevention (CDC) estimated that at least 30% of human antibiotic prescriptions were unnecessary. In 2019, the CDC found that each year 2.8 million Americans have serious infections caused by antibiotic-resistant bacteria. These infections result in about 35,000 deaths annually. Clearly, antibiotics are in danger of becoming victims of their own success.

FURTHER INFORMATION

Aminov, Rustam I. "A Brief History of the Antibiotic Era: Lessons Learned and Challenges for the Future." *Frontiers in Microbiology* 1 (December 8, 2010): 134. https://www.ncbi.nlm.nih.gov/pmc/articles/PMC3109405

BI Biographics. "Alexander Fleming: The Father of Antibiotics." YouTube, December 21, 2019. https://www.youtube.com/watch?v=nh9sHtJrL9s

Macfarlane, Gwyn. *Alexander Fleming, the Man and the Myth.* New York: Oxford University Press, 1985.

35. First Gender Affirmation Surgery, 1930

The first gender affirmation surgery, formerly called sex reassignment surgery, was performed on a Danish citizen born Einar Wegener (1882–1931) who died with the legal name Lili Ilse Elvenes but was best known as Lili Elbe. Wegener was considered male at birth. She presented as a male for many years before she transitioned to a woman, but according to her posthumously published autobiography, *Man into Woman*, her gender identity, or internal image of herself, was always strongly female. Today this condition is designated as gender dysphoria.

Wegener studied art at the Royal Danish Academy for Fine Arts, and by the time she was 25, she was winning prizes for her landscape paintings. While at the Academy, still anatomically male and presenting as such, she met Gerda Gottlieb (1886–1940), another art student. They married in 1904. Gottlieb was ultimately a more successful artist than Wegener, who gave up painting when she began to present as Lili Elbe. Gottlieb worked as an illustrator for fashion magazines and became known for her paintings of beautiful women, some of whom were Wegener in women's clothes.

Wegener's transition to Lili Elbe began when one of Gottlieb's models failed to show up for a session and Gottlieb asked Wegener to put on women's clothes and fill in for the model. At that point, Wegener felt connected to female gender identity, and the transition to Lili Elbe began. She continued to model for Gottlieb and soon began to present as Lili in public, often being introduced as Wegener's sister. Apparently Wegener had a slight build and delicate features, because she claimed that she was harassed in Paris when she wore male clothes and was accused of being a woman trying to pass herself off as a man. This dual identity continued for the next 20 years, but by 1930, it was not enough. Lili Elbe wanted to be physically female as well as socially accepted as a woman. The stress of her conflicted life made her decide that if she could not physically transition, she would kill herself on May 1, 1930.

In her quest to surgically become fully female, Elbe visited Magnus Hirschfeld (1868–1935), a German physician and sexologist. He was a closeted gay man and was particularly interested in ending legal and social discrimination against gay men, although he also studied nonbinary variations of gender identity and sexual orientation. In 1919, Hirschfeld established the Institute for Sexual Science in Berlin. He hoped to find scientific proof that sexual identity differences were biological in origin rather than arising from psychological abnormalities, which was the dominant thinking at the time.

The Institute attracted people from across Europe who had alternative sexual orientations and gender identities. Elbe visited in 1930 in her quest to physically become a woman. After evaluating Elbe, Hirschfeld became a consultant for her surgeries. These surgeries were highly experimental and quite risky in the preantibiotic era.

The first surgery was performed in 1930 by Dr. Erwin Gohrbandt (1890–1965), who castrated the 48-year-old anatomically male Elbe by removing her testicles. After that, Elbe declared she was a woman, and her remaining three or four (the number is unclear) surgeries were performed at the Dresden National Women's Clinic by Dr. Kurt Warnekros (1882–1949). First, Warnekros removed her penis. In the next operation, he created a vagina. Although the records of the clinic were destroyed during World War II, it appears that in a third operation he implanted ovaries removed from a 26-year-old woman at the clinic. A final operation involved transplanting a uterus in the hope that Elbe could have children. This operation ultimately proved fatal. Lili Elbe died three months later, most likely because her body rejected the transplanted uterus.

Transitioning from male to female caused some legal issues. In 1904, Einar Wegener, assigned male at birth, married Gerda Gottlieb, assigned female at birth. After Wegener physically transitioned to Lili Elbe, Gerda Gottlieb tried to obtain a divorce. However, the court ruled that since Lili Elbe and Gerda Gottlieb were both female, there could have been no marriage, and thus there could be no divorce. It took a special act of King Christian X of Schleswig-Holstein-Sonderburg-Glücksburg to annul the marriage in October 1930 so that Gottlieb could remarry. At the time of the divorce, the former Einar Wegener was issued a passport in the name of Lili Ilse Elvenes. However, Lili Elvenes was known for years—both before and after the legal name change—only as Lili Elbe. A film, The Danish Girl, based on her life was released in 2015.

HISTORICAL CONTEXT AND IMPACT

There have always been people who do not fit society's binary model of female (XX chromosomes, female gender identity, and sexually attracted to males) or male (XY chromosomes, male gender identity, and sexually attracted to females). Throughout history, people who did not fit society's concept of male or female have been considered deviants and been persecuted, incarcerated, and even put to death. Their response was often to live secret lives full of desperation and stress that sometimes resulted in self-mutilation or suicide.

Over time, various explanations have been given for gender nonconforming individuals. The most common explanation has been that the individual was mentally ill and needed to be "cured." This resulted in everything from involuntary committal to mental institutions to electric shock treatments to forced castration. Another common explanation has been that nonconforming gender resulted from the way children were socialized, including the clothes they wore, the toys they played with, and the parent they identified with. This does not make much sense when one considers that in the late 1800s and early 1900s little boys wore dresses until they went to school and pink was considered a "boys' color,"

Lili Elbe, the first documented person to have gender affirming surgery, pictured here after she transitioned from male to female. (Wellcome Collection. Attribution 4.0 International (CC BY 4.0))

yet this social trend did not result in several generations of widespread gender nonconformity. With a better understanding of hormones and genetics, many researchers now favor an explanation that includes a natural variation or continuum of biological and hormonal influences.

Because gender nonconformity was, and still is, illegal and punishable in many countries, gender dysphoria has been difficult to study. Hirschfeld was a pioneer in the field. He had the advantage of living in Prussia (now part of Germany) at a time when laws against homosexuality were not enforced. This ended in 1933 when the Nazis shut down the Institute for Sexual Science and publicly burned Hirschfeld's work. Hirschfeld was in France at the time and remained there until his death. Serious transgender research did not begin again in Germany until the 1960s. In the years after World War II, Denmark had the most active and accepted gender affirmation program, although Danish doctors were not allowed to perform this surgery on people who were not Danish citizens.

In the United States, Harry Benjamin (1885–1986), a student of Hirschfeld's, worked with transgender people in the late 1940s in San Francisco, as did Robert Stoller (1924–1991) at the UCLA Gender Identity Clinic. In 1952, Christine Jorgensen (1926–1989) became the first American publicly known to have had gender affirmation surgery. At the time, her case was widely sensationalized in newspapers.

Despite the work of Benjamin and Stoller and the pioneering work of sexologist Alfred Kinsey (1894–1956), acceptance of transgender and nonconforming gender identity has been slow to come in the United States, where nonconforming gender was often considered a mental illness. Finally in 2013, the *Diagnostic and Statistical Manual of Mental Disorders* (DSM–5), used by the American Psychiatric Society for defining the diagnostic criteria of mental illnesses, explicitly stated that "gender non-conformity is not in itself a mental disorder." However, the publication noted that distress caused by gender dysphoria may cause other mental illnesses.

In 2021, about 1.4 million Americans or 0.6% of the U.S. population self-identified as transgender. Not all trans people have full gender affirmation surgery. Some choose to modify their bodies through the use of hormones and/or less extensive surgeries. Most studies have found that 90% or more of people who undergo gender affirmation surgery are pleased with the results, especially if they work with a therapist before and after their transition.

Acceptance of and legal protections for trans people vary tremendously in different countries. Although some modern societies are more accepting of and offer legal protections to trans people, the threat of criminalization remains, especially in some Muslim and African nations. Thailand performs the largest number of gender affirmation surgeries, mostly on foreigners. Iran performs the second largest number of such surgeries. Homosexuality is punishable by death in Iran. In order to rid the country of homosexuality, the Iranian government will pay for gender affirmation surgery even though sexual orientation and gender identity are two separate things and even though most gay men do not self-identify as female.

FURTHER INFORMATION

"About Transgender People." National Center for Transgender Equality, Undated. https://transequality.org/about-transgender

Clay, Rebecca. "Embracing a Gender-Affirmative Model for Transgender Youth." *American Psychological Association Monitor on Psychology* 49, no. 8 (September 2018): 29. https://www.apa.org/monitor/2018/09/ce-corner

Cox, David. "The Danish Girl and the Sexologist: A Story of Sexual Pioneers." *The Guardian*, January 13, 2916. https://www.theguardian.com/science/blog/2016/jan/13/magnus-hirschfeld-groundbreaking-sexologist-the-danish-girl-lili-elbe

Elbe, Lili, with Niels Hoyer, ed. *Man into Woman: An Authentic Record of a Change of Sex.* New York: E. P. Dutton, 1933. Re-released for Kindle in 2015 under the name *Lili: A Portrait of the First Sex Change.*

36. First Successful Joint Replacement, 1940

The first successful joint replacement was a hip replacement performed by Austin Talley Moore (1899–1963) in conjunction with surgeon Harold Bohlman (1893–1979). The surgery was done under spinal anesthesia at Johns Hopkins Hospital in Baltimore, Maryland, on September 28, 1940.

Austin Moore was born in Ridgeway, South Carolina, and graduated from the Medical College of South Carolina in 1924. After furthering his education with a residency under A. Bruce Gill, a renowned orthopedist at the University of Pennsylvania, he returned to South Carolina in 1928 to found the Moore Orthopedic Clinic. Harold Bohlman was born in Iowa, served in the U.S. Air Service during World War I, and then graduated from Johns Hopkins Medical School in 1923. Bohlman was an orthopedist and trauma surgeon with a special interest in knee and hip surgery.

The hip is a ball-and-socket joint. The rounded head of the femur (thigh bone) fits into the acetabulum, a socket that is formed by bones of the pelvis. The acetabulum is lined with cartilage. When the hip joint works correctly, the head of the femur moves smoothly over the cartilage within the acetabulum allowing pain-free movement. Older individuals often experience hip pain from arthritis and wearing away of the cartilage, but the hip joint can also be damaged by inflammation, disease, or bone fracture. Damage can make walking painful at best and impossible at worst.

There are two types of hip replacements—a hemiarthroplasty, in which only one part of the joint is replaced, and a total hip arthroplasty (THA), in which both the femur head and acetabulum are replaced with a prothesis. L. C. Clarke (1894–1942) was the first person to have a successful hip replacement operation. He had a hemiarthroplasty—only the femur head was replaced. This operation is still done today, although THA is much more common.

Clarke, at age 46, was a large man, weighing about 275 pounds (125 kg). For several years, he lived with a fast-growing giant cell tumor near the femur head. The tumor was not cancerous but was extremely painful. It was the size of a grapefruit and interfered with the movement of his hip. The tumor had been surgically removed twice, but both times it quickly grew back. To complicate matters, Clarke had a history of fractures of the femur neck, the area immediately below the femur head. After examination, Moore concluded that the only treatment choices left for Clarke were to amputate at the hip joint, which would cause permanent disability, or to try the unproven procedure of replacing the femur head with a manufactured prothesis.

An effective prothesis needed to be able to bear Clarke's weight, and it had to be large enough not to be dislodged from the acetabulum during movement yet move smoothly within it. Most importantly, it had to be made of material that would not crack, break down, corrode, or interact with the body. Harold Bohlman had served in the U.S. Air Service. He was aware that airplane parts had some of the same requirements as a prothesis for the body. They had to withstand harsh conditions and high temperatures without corroding, cracking, or breaking. This problem in airplanes was solved by making some parts from a metal alloy called vitallium, which was developed in 1932. Vitallium is composed of cobalt, chromium, and molybdenum. Bohlman believed it would not corrode, break, flake, or harm the body.

Bohlman suggested that the femur head and neck, where the tumor grew, should be cut out and replaced with a vitallium prothesis. Using measurements from X-rays, the doctors enlisted Austenal Laboratories, a maker of airplane parts, to produce a 12-inch (30 cm) prothesis with a ball head and two prongs. The ball head would fit in the acetabulum, which was left intact. The prongs would slide over the cut end of the femur. The prothesis also had metal loops on the sides so that muscles previously attached to the femur could be attached to the prothesis.

Moore recognized that the operation was groundbreaking and hired a professional cameraman to record the event. The operation did not go smoothly. Clarke bled heavily, and when the time came to slip the prongs of the prothesis over the cut femur, the femur was too wide to fit. The surgeons had to chip away at the outer edges of the bone to make it thin enough to let the prongs slide into place. Then, a few days after the operation, the femur below the lower end of the prothesis fractured. By using traction and extended bed rest, the ends of the break were kept in alignment until the bone healed.

Despite these setbacks, nine months after the operation, Clarke could walk well without artificial support, although he occasionally used a cane when walking long distances. He lived another two years and then died of congestive heart failure, a condition unrelated to his hip surgery. Moore removed the prothesis at the autopsy and found no signs that it had corroded, broken down, or harmed any body tissue. The giant cell tumor had not returned, and the acetabulum was disease free. Moore was the first physician to have a joint replacement patient recover the ability to walk unaided and to have the prothesis remain intact until the patient's death.

HISTORICAL CONTEXT AND IMPACT

The earliest recorded attempt at joint replacement was made by Themistocles Gluck (1853–1942). Gluck was born in Moldavia (now part of Romania), studied under a number of well-known physicians in Berlin, and later practiced medicine in that city. He also served as a surgeon during war in the Balkans, where his interest in bone surgery likely developed. On May 20, 1890, he performed his first joint replacement surgery using an ivory prothesis to replace part of the knee. He chose ivory because, of the options available, it seemed most bonelike to him. In 1891, he used an ivory ball-and-socket joint in a hip replacement surgery. Initially both operations were successful, but the ivory soon degraded and the joint failed.

The key to successful joint replacement was to find a material that stood up to heavy use of the joint and still allowed the femur to move smoothly. Varied materials and combinations of materials were tried. In 1925, Norwegian-born American surgeon Marius Nygaard Smith-Petersen (1886–1953) made a glass hemisphere that fit into the acetabulum with the idea that glass would provide a low-friction surface for the prothesis to move against. Unfortunately, glass was not strong enough to survive the stress put on it, and it shattered. In 1938, Jean Judet (1913–1995) and Robert Judet (1909–1980), a pair of Parisian doctor brothers, tried lining the acetabulum with an acrylic substance. The acrylic provided a smooth surface, but it soon separated from the acetabulum causing failure of the joint. Vitallium was the first material that created a prothesis that met the requirements of strength and durability.

The quest for improved materials continued after Moore and Bolhman's success. British surgeon John Charnley (1911–1982) is considered the father of modern THA. He approached the problem of hip joint replacement from a biomechanical perspective, concentrating on the effects of bone compression and friction. He used a metal implant with a spike that fit into the center of the femur for strength, and he lined the acetabulum with Teflon to reduce friction. Everything was fixed in place with dental cement. Unfortunately, Teflon did not hold up well and also interacted with soft tissue. He replaced Teflon with high-molecular-weight polyethylene, which created a safe low-friction surface and ensured durability.

Other researchers tried metal-on-metal protheses, but these could loosen and had the potential to cause cancer. Ceramic-on-ceramic protheses, although providing a low-friction surface, proved expensive, difficult to implant, and sometimes made noise with each step the patient took.

Today improved variations of Charnley's model are still in use. Surgeons have also experimented with minimally invasive THA that requires only about a four-inch (10 cm) incision. Robot-assisted surgery is also becoming more common. It allows for more exact placement of the prothesis. According to the American Academy of Orthopaedic Surgeons, hip and knee replacements are the most common joint replacements. An estimated 238,000 THAs were performed in the United States in 2020. The number was expected to grow to 652,000 in 2025 and to 850,000 in 2030 as the American population ages.

FURTHER INFORMATION

Chillag, Kim J. "Giants of Orthopaedic Surgery: Austin T. Moore MD." *Clinical Orthopaedics and Related Research* 474, no. 12 (December 2016): 2606–2610. https://www.ncbi .nlm.nih.gov/pmc/articles/PMC5085962

DePuy Synthes Companies. "The Latest Procedure: Anterior Approach Total Hip Replacement." YouTube, September 8, 2014. https://www.youtube.com/watch?v=5NqJa_J2dfw

Heringou, Philippe, Steffen Quiennec, and Isaac Guissou. "Hip Hemiarthroplasty: From
    Venable and Bohlman to Moore and Thompson." *International Orthopaedics* 38, no.
    3 (March 2014): 655–661. https://www.ncbi.nlm.nih.gov/pmc/articles/PMC3936081
Knight, Stephen R., Randeep Anjula, and Satya P. Biswas. "Total Hip Arthroplasty—
    Over 100 Years of Operative History." *Orthopedic Reviews* 3, no. 2 (September 6, 2011):
    e16. https://www.ncbi.nlm.nih.gov/pmc/articles/PMC3257425

37. First City with Fluoridated Water, 1945

On January 25, 1945, Grand Rapids, Michigan, became the first city in the world to have artificially fluoridated municipal water. Fluoride is the 13th most common mineral on Earth. It naturally leaches into water from rocks and soil. It ionizes in water and is then called fluorine. Because of the soil composition, some locations have naturally fluoridated water. Grand Rapids does not. The city's decision to fluoridate its water supply came after more than 30 years of research, the results of which convinced the city administration that fluoridation would improve health and save residents money by decreasing tooth decay.

The story of fluoridation begins in 1901 with dentist Frederick S. McKay (1874–1959). McKay was born in Lawrence, Massachusetts, and educated in local public schools. His dream was to become a musician, but he developed a suspected case of tuberculosis and, in 1894, moved to Colorado where the climate was better for his health. Within three years, he was back in Massachusetts working as a trolley conductor. Encouraged by his brother-in-law, he enrolled in Boston Dental College (now Tufts University Dental School), later transferred to the University of Pennsylvania, and graduated in 1900.

By 1901, McKay was back in Colorado working as a dentist. He was disturbed by the large number of children around Colorado Springs who had permanent chocolate-colored stains on their teeth. Locally this condition was called Colorado brown stain. Area residents had many theories about what caused the stain, including eating too much of certain foods, radium exposure, and a deficit of calcium in drinking water, but McKay was the only local dentist interested in scientifically investigating the cause.

McKay wrote to Greene V. Black (1836–1915), dean of the dental school at Northwestern University, about Colorado brown stain. In 1909, Black was interested enough to travel from Chicago to Colorado Springs to study the discoloration. The men received several small grants and collaborated on brown stain research until Black died in 1915. By that time, they had concluded that something in the local water caused tooth discoloration, but their equipment was not good enough to identify the substance. During their examinations of many children who had "mottled enamel," their new name for Colorado brown stain, they observed that these children had healthier, more decay-resistant teeth than children without tooth discoloration.

The next clue to the origin of the brown stain came from Bauxite, Arkansas, where the Aluminum Company of America (ALCOA) had a large mine. Bauxite's children also had mottled enamel teeth. Harry V. Churchill (1887–), ALCOA's chief chemist, did not care about tooth discoloration, but he was concerned about rumors that aluminum cooking pots were toxic. He thought the mottled enamel teeth of Bauxite children might be used as evidence to support these rumors, so he ordered an analysis of Bauxite's water. Repeated analyses showed that the water contained an unusually high level of fluorine. Churchill happened to have read McKay's reports on mottled enamel. He wrote McKay in Colorado and asked him to send water samples from towns where people had the brown staining. Every water sample McKay sent showed a high level of fluorine.

The mystery of the brown stains and discolored enamel had been solved. The cause was excessive fluorine in the water. One question still remained. Why did children with mottled tooth enamel have fewer cavities and less tooth loss than children without brown stained teeth? H. Trendley Dean (1894–1962), head of the Dental Hygiene Unit at the National Institute of Health, wanted to know the answer. He arranged for funding of various investigations that showed that drinking water containing up to 1.0 parts per million (1 mg/L) of fluorine was protective against tooth decay.

Tooth enamel is the strongest substance in the body, even stronger than bone. It is made of a compound of calcium and phosphate. When bacteria in the mouth feed on traces of sugars left from food, they produce an acid that erodes tooth enamel. Calcium and phosphate in saliva can repair the enamel, but when enamel loss occurs faster than repair, cavities, infection, pain, and tooth loss will occur. The body chemically incorporates fluorine into tooth enamel. This makes the enamel stronger and more resistant to destruction. On the other hand, too much fluorine results in strong but discolored enamel, a condition now called fluorosis.

After many discussions with the U.S. Public Health Service and the Michigan Department of Health, the city commissioners in Grand Rapids voted to fluoridate their water in 1945 to improve residents' dental health. Over the next 11 years, dentists examined more than 30,000 Grand Rapids school children and found that their rate of tooth decay had decreased by 60%. Other studies in the United States, Canada, and the United Kingdom compared towns of equal size with and without artificially fluoridated water. The researchers found substantially less tooth decay in every town using fluoridated water. Not only had the mystery of Colorado brown stain been solved, the solution offered a new tool for improving public health.

HISTORICAL CONTEXT AND IMPACT

Fluoridation of water has been declared one of the 10 greatest public health advances of the twentieth century. The benefit extends across all income levels and ethnicities to such a degree that today we do not even think of tooth decay as a public health problem. Things were different before fluoridation. Dental researchers Hilleboe and Ast reported in 1951 that men between the ages of 20 and 35 had already lost an average of 4.2 teeth, had on average one tooth that needed extracting, 7.2 surfaces that needed fillings, and 90% of these men needed bridges or partial or full dentures.

Although it has been known for 75 years that fluoridation can significantly prevent tooth decay, the United States has no federal mandate requiring water to be fluoridated. In 2011, the recommendation for fluoridated water was lowered from 1.0 parts per million (1.0 mg/L) to 0.7 ppm (0.7 mg/L) with adjustments based on local conditions. This level has been determined to be protective without causing significant tooth discoloration.

Fluoridating the water supply has not come without a fight. During the 1950s at the height of the Cold War, the idea circulated that fluoridation was part of a Communist plot to destroy America. Local fights about whether to fluoridate have led to repeated votes, repeals of decisions, violence, and death threats. Since 1956, Portland, Oregon, for example, has voted at least five times on the question of fluoridation. Sometimes the referendum failed. Other times, it passed, only to later be challenged and repealed. As of 2022, Portland remains the largest American city without fluoridated water.

About three-quarters of Americans using public water supplies receive fluoridated water, but some anti-fluoride groups still remain active. These opponents believe that fluoridating water infringes on people's right to determine what goes into their body. They also argue that fluoridation is no longer necessary. Starting in the 1970s, fluoride has been incorporated into most toothpastes and mouthwashes in the United States. Other countries have fluoridated salt or milk. Anti-fluoride groups feel people who want fluoride can get enough from these sources. Although the proportion of fluoride people get from water and beverages made with fluoridated water has decreased with the increased use of fluoridated toothpaste, research shows that fluoride from non-water sources alone is inadequate to protect teeth.

Anti-fluoride activists have also claimed that fluoridation causes health problems ranging from cancers to damage to skeletal bones, joint pain, seizures, neurological problems, high blood pressure, and acne. No well-controlled scientific studies have substantiated these claims. Court cases have also been filed stating that the effects of fluoridation on pregnant women and fetuses have not been adequately investigated. A court case based on alleged violations of the 1976 Toxic Substances Control Act was filed in 2016 and is still unresolved as of 2022. The suit claims that fluoridation lowers children's IQs.

One huge impact of fluoridation is acceptance of the idea that teeth can be protected from decay. This has had the effect of changing dentistry from reactive care to preventative care. Before fluoridation, people tended to visit a dentist only when they could not stand a toothache any longer, and the common solution was to extract the tooth. Once the public embraced the idea of preventing tooth decay, the view of dentists changed. Regular dental checkups came to be viewed as a normal and necessary part of preventative care.

FURTHER INFORMATION

Carstairs, Catherine. "Debating Water Fluoridation Before Dr. Strangelove." *American Journal of Public Health* 105, no. 8 (August 2015): 1559–1569.

Hilleboe, Herman E., and David B. Ast. "Public Health Aspects of Water Fluoridation." *American Journal of Public Health* 41, no. 11, pt. 1 (1951): 1370–1374.

LG News. "Why the Government Puts Fluoride in Our Water." YouTube, February 19, 2015. https://www.youtube.com/watch?v=XuMxAB9q92E

Mullen, Joe. "History of Water Fluoridation." *British Dental Journal* 119, no. S7 (2005): 1–4. https://www.nature.com/articles/4812863

"Timeline for Community Water Fluoridation." Centers for Disease Control and Prevention, April 28, 2021. https://www.cdc.gov/fluoridation/basics/timeline.html

38. First Woman to Win the Nobel Prize in Physiology or Medicine, 1947

In 1947, biochemist Gerty Cori (1896–1957) became the first woman to win the Nobel Prize in Physiology or Medicine and only the third woman to win in any of the sciences. She shared the prize with her husband, Carl Ferdinand Cori (1896–1984), and physiologist Bernardo Houssay (1887–1971), who was the first Argentine to win a Nobel Prize.

Gerty Theresa Cori (Gerty was her full name, not a nickname) was born Gerty Radnitz in Prague in 1896. At that time, Prague was part of the Austro-Hungarian empire. Today it is the capital of the Czech Republic. Gerty Radnitz was the oldest of three girls in a Jewish family. The family was well educated. Otto Radnitz, her father, was a chemist. He developed a method for refining sugar and had a career managing sugar refineries. Her mother was educated and well read at a time when many women received little advanced education.

Gerty was educated at home until she was 10. Later, she attended a private school for girls. She wanted to attend medical school, but her school did not teach the prerequisite classes in the sciences and mathematics. Nevertheless, with the support of her family and intense supplementary study, in 1914, she met the entry requirements to enroll in medical school at the German Karl-Ferdinands-Universität in Prague.

During Gerty's first year as a medical student, she met her future husband, Carl Cori, in anatomy class. Carl came from a Catholic family of scientists. He grew up in Trieste, now part of Italy, where his father was the director of a marine biology station. The couple shared a love of science and outdoor activities, especially skiing and mountain climbing.

Carl was drafted into the Austro-Hungarian army in 1916 and served until the end of World War I in 1918, after which he returned to medical school. Both Gerty and Carl received their medical degrees in 1920. Gerty then converted to Catholicism and married Carl.

The couple moved to Vienna where a pattern started that would continue for most of their professional careers. Carl obtained a good position at the University of Vienna, while Gerty, who had the same education as Carl, could find work only as a low-level assistant pathologist because she was a woman. Carl was invited to work with future Nobel Prize winner Otto Loewi (1873–1961) in Graz, where his lifelong interest in sugar metabolism began. Gerty stayed stuck in Vienna with her low-level job.

Living conditions after the war were difficult. At one point, Carl could only continue his research on sugar metabolism because his father sent him frogs from the marine biology station in Trieste. Gerty's diet in Vienna was so limited that she developed the eye disease xeropthalmia, which is caused by vitamin A deficiency.

Antisemitism was rising across Europe. Gerty was still considered a Jew despite converting to Catholicism. Between being Jewish and a woman, she had little hope of finding a research position, and being married to a Jew could derail Carl's career too. The couple decided to move either to the United States or to the Dutch island of Java where they could practice medicine. A job offer for Carl from the State Institute for the Study of Malignant Disease (now called Roswell Park Comprehensive Cancer Center) in Buffalo, New York, came through first, so in 1922 Carl moved to Buffalo. Gerty followed six months later.

Once again, Carl was offered a good research job, but Gerty was only grudgingly given a job as an assistant pathologist. Having grown up with a father who researched sugar and a grandfather who developed diabetes, she was, like Carl, interested in sugar metabolism and kept wandering into Carl's lab to work with him. This happened so often that the director of the Institute threatened to fire her if she did not confine her work to the pathology lab. Once again, she persisted and eventually was allowed to work with Carl. Together they researched glucose metabolism in cancer tumors. Their continued research in glucose metabolism would eventually lead to the Nobel Prize in 1947.

The Coris became naturalized U.S. citizens in 1928. After publishing more than 50 papers together while in Buffalo, Carl was offered a job at several universities but refused because the schools would not also hire Gerty. In 1931, Washington University School of Medicine in St. Louis, Missouri, agreed to make an exception to their rule that married couples could not work together. Carl accepted the job of chairman of the pharmacology department. Gerty was hired as a research associate at a salary one-tenth of what Carl was paid. She was not promoted to full professor until she won the Nobel Prize.

The Coris were awarded the Nobel Prize for discovering what is now called the Cori cycle or the lactic acid cycle. Glucose is a simple sugar that fuels every cell in the body. It is stored in muscles and in the liver as glycogen, a more complex molecule. When muscles work hard, glycogen is broken down into glucose for energy. When muscles use more oxygen than they have available, a byproduct of this breakdown is lactic acid. Lactic acid is what causes "muscle burn" and fatigue during intense exertion.

The Coris discovered that lactate from muscle activity is released into the bloodstream and carried to the liver. Here another set of chemical reactions converts lactate into glycogen that can be broken down into glucose and released back into the bloodstream where it travels to muscle and other cells to replenish the glucose used in muscular activity. This cycle is driven by a number of complex chemical reactions and enzymes that the Coris discovered and isolated.

Gerty Cori remained at Washington University for the remainder of her career and continued to make discoveries related to the regulation of sugar metabolism. She and Carl had one child, Tom (1936–), who is interviewed in the PBS YouTube video referenced in the Further Information section of this entry. About the time the Coris won the Nobel Prize, Gerty Cori developed myelosclerosis, a disease in which bone marrow becomes fibrous and can no longer produce an adequate supply of blood cells. She died in 1957 and donated her body to science. Carl remarried in 1960 and died in 1984.

HISTORICAL CONTEXT AND IMPACT

Gerty Cori's achievements were important on two levels—social and scientific. During her lifetime, women were not expected to be educated in the sciences or to do original scientific research. Although Carl considered Gerty a full research partner, his employers did not. If Carl had not made it clear that if they wanted him to work for them they would have to hire Gerty, it seems unlikely that she would ever have been considered for the Nobel Prize.

Employers did hire Gerty, but only grudgingly. She was always assigned low-status assistant jobs at significantly less pay than Carl rather than being recognized as a top-notch researcher in her own right. Even the Nobel Prize committee appears to have considered her an extension of Carl rather than a full partner in the research that came out of their laboratory. Instead of dividing the money that accompanies the Nobel Prize equally among the three recipients, Gerty and Carl

Gerty Cori, the first woman to win the Nobel Prize for Physiology or Medicine (1947) for her work on glucose metabolism. (Courtesy National Library of Medicine)

each received one-quarter of the money while Houssay, who researched a different metabolic problem, received one-half. Although Gerty Cory broke many gender barriers, even after she became a Nobel laureate, she was excluded from multiple prestigious prizes that were awarded only to Carl. This bias against women in the sciences continues today.

As of 2022, only 12 women have been awarded the Nobel Prize in physiology or medicine, 8 in chemistry, and 4 in physics, out of 636 laureates in the sciences. Marie Curie (1867–1934) was the first woman to win a Nobel Prize in the sciences. She won in physics in 1903 with her husband and was the single winner in chemistry in 1911. In 1922, Cécile Vogt-Mugnier (1875–1962) became the first woman to be nominated for a Nobel Prize in Physiology or Medicine for her examination of the structure and function of the brain. She did not win. Neither did Maud Slye (1879–1954), a pathologist who was nominated in 1923. After Gerty Cori's 1947 win, 30 years passed before another woman, Rosalyn Yallow (1921–2011), won the Prize in medicine. Between 1978 and 2023, only 10 more women won the Prize in physiology or medicine, most recently Tu Youyou (1930–) from China in 2015.

In addition to breaking social barriers as a woman, Gerty Cori's work was significant on a scientific level. Insulin, which regulates the uptake of glucose into cells, was not isolated until 1921 (see entry 33). Before this, people who developed diabetes, an insulin deficiency disorder, died. Gerty's insight and work in determining the mechanisms of sugar metabolism resulted in a better understanding of diabetes. Based on the understanding of how glucose is regulated, multiple classes of drugs have been developed to treat type 2 diabetes, the most common form of the disorder. Each class of drugs affects glucose metabolism in a different way to meet the needs of individual patients, making diabetes treatment more effective.

FURTHER INFORMATION

The Movement System. "What Is the Cori Cycle? Gluconeogenesis Explained Simply." YouTube, September 30, 2020. https://www.youtube.com/watch?v=IFns6KaHzEA

National Historic Chemical Landmarks of the American Chemical Association. "Carl and Gerty Cori and Carbohydrate Metabolism." September 24, 2004. http://www.acs .org/content/acs/en/education/whatischemistry/landmarks/carbohydratemetabolism .html

NobelPrizeOutreach. "GertyCori—Biographical."December1947.https://www.nobelprize .org/prizes/medicine/1947/cori-gt/biographical

PBS Nine. "The Coris: Living in St. Louis." YouTube, November 28, 2007. https://www .youtube.com/watch?v=GvaC2XCjGgw

39. First FDA-Approved Chemotherapy Drug, 1949

The first effective chemotherapy drug to treat cancer was developed from an agent of chemical warfare called mustard gas. This gas killed over 91,000 people and injured another 1.2 million in World War I (1914–1918). Mustard gas is laboratory-made; it does not exist in nature. The gas was first synthesized, and its toxic effects noted, in 1860 by British physicist Frederick Guthrie (1833–1886) and almost simultaneously by German chemist Albert Niemann (1834–1861), whose death is believed to have been caused by exposure to the gas. Although this chemical is called a gas, at room temperature it is a liquid. To be an effective chemical weapon, it must be spread in the air as fine particles, usually through an explosion.

The Germans first used mustard gas in France in July 1917, but it was used by both sides during World War I. Mustard gas causes blindness and blisters over any part of the body it touches. If inhaled, blistering in the lungs can be fatal or cause lifelong breathing difficulties. In 1919, Edward Krumbhaar (1882–1966), an American military medical officer, examined soldiers injured by mustard gas and reported that they all showed a severe drop in the number of white blood cells.

Use of chemical and biological weapons in warfare, including mustard gas, was banned by the Geneva Protocol in 1925. Despite signing the Protocol, the U.S. Congress did not ratify the treaty to make it binding until 1975, when it also rati-fied the Biological and Toxic Weapons Convention. Thus, in response to World War II (1939–1945), the U.S. Office of Scientific Research and Development began a secret program at Yale University to investigate mustard gas. The program was run by pharmacologists Louis S. Goodman (1906–2000) and Alfred Z. Gilman (1908–1984).

As part of their research, Goodman and Gilman followed up on Krumbhaar's findings in World War I that mustard gas affected white blood cell production. On December 2, 1943, an act of war accelerated their interest in mustard gas as a therapeutic agent. On that date, the American Liberty ship *John Harvey* was anchored in the harbor at Bari, Italy. The ship carried 100 tons of mustard gas. When the German Luftwaffe bombed the harbor, it destroyed the *John Harvey*. Mustard gas was released on soldiers stationed in Bari and on unsuspecting townspeople. Soon, many became ill and some died. Lieutenant Colonel Stewart F. Alexander (1914–1991), a physician with special training in chemical warfare, examined those who had been exposed and noted that they all had exceptionally low white blood cell counts. His report added support for continuing the therapeutic research Goodman and Gilman were doing at Yale.

Blood cells are formed in bone marrow. White blood cells are important in fighting disease because they can divide rapidly, but this rapid division is a liability when the cells become abnormal or cancerous. Leukemias are cancers of the bone marrow in which large numbers of immature or abnormal white blood cells are formed. Lymphomas are cancers of the lymph nodes in which white blood cells develop a mutation that causes them to multiply rapidly and increases the number of cells that are cancerous. Because exposure to mustard gas reduces the number of white blood cells that the bone marrow produces, it was investigated as a potentially effective agent against leukemias and lymphomas.

Through experimenting on mice, Goodman and Gilman discovered that mustard gas damages DNA in a way that prevents both healthy and cancerous white blood cells from reproducing. Because abnormal white blood cells reproduce faster than healthy white blood cells, they hoped to use mustard gas to kill off many abnormal cells without damaging too many healthy cells.

Mustard gas consists of carbon, hydrogen, chlorine, and sulfur. It is highly toxic and difficult to handle. For therapeutic research purposes, Goodman and Gilman removed the sulfur atom and added a nitrogen atom. This new compound had the same effect on blood cell reproduction but was easier to handle. Their research was so secretive that in their notes they called the compound "X." Later, it became known as nitrogen mustard.

On August 27, 1942, Goodman and Gilman treated their first patient with nitrogen mustard. The man was identified only as JD, a 47-year-old Polish immigrant. He was a patient of Yale surgeon Gustaf Lindskog (1903–2002), who had diagnosed him as having an advanced inoperable lymphoma with no chance of survival. Because Goodman and Gilman had no idea what dose of nitrogen mustard to give the patient, they gave him 10 doses of varying strengths over 10 days. Almost miraculously, his tumor shrank, but the results were not permanent, and JD died in December. In 1949, after additional research, the drug, now named mechlorethamine became the first FDA-approved chemotherapy drug. Today it goes by the trade name Mustargen and is still used to treat some cancers.

Neither Gilman nor Goodman had careers in cancer research. Gilman founded the department of pharmacology at Albert Einstein College of Medicine. Goodman founded the department of pharmacology at the University of Utah College of Medicine. Together they wrote a definitive textbook on pharmacology, which was in its 14th edition in 2022.

HISTORICAL CONTEXT AND IMPACT

Although for years medical charlatans had peddled ineffective quack cures for cancer, surgery was its only legitimate medical treatment until the development of mechlorethamine. For most patients, however, the reality was that by the time their cancer was diagnosed, it was too late for surgery to be effective. In addition, surgery was not possible for many cancers, such as leukemias. Even when surgery was an option, operations carried high risks of infection, complications, and failure. The development of mechlorethamine gave researchers hope that a drug cure for cancer could be found.

Initial efforts to find a drug cure involved looking for compounds that were similar to mechlorethamine. This was unsuccessful "try and hope" research because of the limited understanding at that time about the metabolic workings of cells. The first drug intentionally designed to treat cancer came into existence because Lucy Wills (1888–1964), an English physician-researcher, discovered that when folic acid, also known as vitamin B9, was given to children with acute lymphocytic leukemia (ALL), it increased the production of ALL cells. Her findings suggested that changing a cell's metabolism could affect cancer. Vitamin B9 was later found to be necessary for DNA synthesis and, along with vitamin B12, is needed for the production of healthy red blood cells.

Sidney Farber (1903–1973), a pathologist at Boston Children's Hospital who studied leukemia in children, was the first person to attempt rational drug design to treat cancer. In 1948, Farber experimentally treated children with ALL using an intentionally designed compound called aminopterin. The drug was designed to interfere with folic acid. Some children Farber treated achieved temporary remission. This inspired a search for other drugs antagonistic to folic acid and led to the development of amethopterin, now called methotrexate. This drug is still widely used to treat many different types of cancer.

In 1955, the U.S. government established the Cancer Chemotherapy National Service Center. Almost all its funding went to drug development. Over time, this initiative produced several somewhat effective anticancer drugs. Many of these early drugs had unpleasant and often damaging side effects. Determining appropriate dosage was difficult, and although some caused remission, there was still no cure for cancer.

By 1970, cancer was the second leading cause of death in the United States. This statistic, along with the continued failure to find a drug cure, stimulated passage of the National Cancer Act of 1971, which President Richard Nixon declared would launch a "war on cancer." The Act provided ongoing funding for basic cancer research, cancer centers, improvements in diagnosis, and drug development. It came at a time when scientists were developing the tools to better understand the mechanisms by which cells became cancerous. This allowed them to begin to develop drugs to slow or halt specific steps in the development of cancer cells, such as blocking the production of enzymes these cells need in order to reproduce. Unfortunately, these drugs also killed healthy cells and often caused substantial side effects, especially when physicians began combining them in multiple drug regimens.

By 1990, cancer prevention education, funding for cancer screenings such as mammograms, and better diagnostic tools resulted in earlier detection of some types of cancer. Early detection and improved treatment resulted in better outcomes, and a steady decline in cancer deaths began, which continues today. In the late twentieth century, researchers also made several breakthroughs in understanding the immune system. With increased knowledge, interest grew in the possibility of using the person's own immune system to fight cancer. Cancer immunotherapy research accelerated.

The first monoclonal antibody treatment for cancer was approved by the U.S. Food and Drug Administration in 1997 (see entry 48). Monoclonal antibodies are laboratory-grown immune system proteins that are engineered to attack specific kinds of cancer cells when injected into the body. In 2010, the first cancer vaccine was approved for use in prolonging the life of men with advanced prostate cancer. Today researchers continue to look for ways in which the immune system can be supported or triggered to attack specific types of cancer cells and for ways to make these individualized cancer treatments affordable.

FURTHER INFORMATION

Conant, Jennet. "The Bombing and the Breakthrough." *Smithsonian Magazine*, September 2020. https://www.smithsonianmag.com/history/bombing-and-breakthrough-180975505

Faguet, Guy B. "A Brief History of Cancer: Age-Old Milestones Underlying Our Current Knowledge Database." *International Journal of Cancer* 136, no. 9 (May 1, 2015): 2022–2036. https://onlinelibrary.wiley.com/doi/full/10.1002/ijc.29134

Godoy, Natalia. "The Birth of Cancer Chemotherapy: Accident and Research." Pan American Health Organization, Undated. https://www3.paho.org/hq/index.php?option=com_content=&view=article=&id=9583=&Itemid=1959=&lang=en

Latosińska, Jolanta Natalia, and Magdalena Latosińska. "Anticancer Drug Discovery—From Serendipity to Rational Design." IntechOpen, January 23, 2013. https://www.intechopen.com/chapters/41943

Montreal Gazette. "Dr. Joe Schwarcz: Mustard Gas and Chemotherapy." YouTube, September 12, 2019. https://www.youtube.com/watch?v=xgrdPuaKBTA

40. First Drug to Treat Depression, 1952

The first antidepressant drug was not intended to treat depression. The drug, called isoniazid, was first synthesized by two Czech scientists in 1912. In 1951, it became a trial drug to treat tuberculosis and through keen observations and chance it became an antidepressant.

In 1951, two doctors, Irving Selikoff (1915–1992) and Edward Robitzek (1912–1984), had a new drug—isoniazid, manufactured by Hoffman-LaRoche. They tested it on patients with tuberculosis at Sea View Hospital, a sanatorium on Staten Island, New York. At the time, no drugs were available to effectively treat the disease. Isoniazid turned out to be a breakthrough drug. It worked so well against tuberculosis that some patients recovered enough to go home and resume normal lives. But the doctors noticed a surprising side effect. Patients who received the drug became more involved in their surroundings. They communicated more and actively engaged with others in ways that they had not before isoniazid treatment. Robitzek reported that some patients became almost euphoric.

Selikoff and Robitzek were not psychiatrists, but they were observant physicians. They included the apparent psychiatric side effects of isoniazid in their reports and journal articles, and then they went on with their tuberculosis research. The doctors also tested a related drug, iproniazid, and found that it had even stronger psychiatric effects. Iproniazid, however, caused liver damage and was withdrawn from the U.S. market in the 1960s.

Neither Robitzek nor Selikoff followed up on the mental and emotional side effects of isoniazid. Edward Robitzek remained at Sea View for 32 years. Irving Selikoff moved on and became a hero of occupational health. After leaving Sea View, he opened a lung clinic in New Jersey. His observational skills again paid off when he noted that 14 of 17 patients who worked in an asbestos factory died from lung cancer, asbestosis, or mesothelioma. Larger studies confirmed the relationship between inhaled asbestos and cancer. The asbestos industry resisted these findings, but Selikoff was persistent and media savvy. By 1989, the Environmental Protection Agency had developed a plan to phase out the use of asbestos. Selikoff's research also resulted in billions of dollars in lawsuits by workers exposed to asbestos.

In the United States, when the Food and Drug Administration (FDA) has approved a drug, it may be advertised only for its approved use, but with a patient's informed consent, physicians can prescribe it for other uses when they feel it would be beneficial. This is known as off-label use. In the early 1950s, there were fewer controls on off-label advertising and prescribing. Two Cincinnati doctors, psychiatrist Max Lurie (1920–2008) and neurologist Harry Salzer (1908–1979), took advantage of off-label use of isoniazid.

Just by chance, Max Lurie read Robitzek's paper on isoniazid treatment for tuberculosis. In an interview near the end of his life, he said that he had no idea why a journal article on tuberculosis would have interested him. Nevertheless, he read the article and was struck by the behavioral side effects Robitzek and Selikoff described. He and Salzer decided to try using the drug to treat patients with major depression. In a 1953 paper, they reported that about two-thirds of the patients who were given isoniazid experienced a significant improvement in depressive symptoms. Max Lurie referred to isoniazid as an "antidepressant," the first use of this word in connection with a drug.

Meanwhile in Paris, psychiatrist Jean Delay (1907–1987) had—also by chance—independently learned of isoniazid's mood-lifting effects from pulmonary physicians at the hospital where he worked. He tried using the drug on his patients who had major depression and reported a similar positive response in a majority of them. Isoniazid had, by observation and chance, become the first pharmaceutical drug to successfully treat depression. At that time, no one understood why the drug worked. Even today, the biochemical mechanism by which isoniazid causes changes in brain function is unclear.

HISTORICAL CONTEXT AND IMPACT

For centuries, opium, amphetamines, and the plant St. John's wort (*Hypericum perforatum*) were used—with little success—to treat what is now called major depressive disorder (MDD). St. John's wort causes mild-to-moderate improvement in depression in a small percentage of people, but it is rarely prescribed by physicians in the United States today because of potentially serious interactions with common drugs such as prescription antidepressants, blood thinners, oral contraceptives, and some HIV and chemotherapy drugs. Until the 1980s, harsh electroconvulsive therapy was also used in the United States to treat MDD and schizophrenia, often with little success.

Using isoniazid to treat depression reinforced the idea that drugs could play a constructive role in treating mental illness, although they worked best when used along with psychotherapy. This set off a hunt for new and better psychotropic drugs. Roland Kuhn (1912–2005), a Swiss psychiatrist, used experimental drugs from the pharmaceutical company Geigy (now Ciba-Geigy) to see if they would relieve psychotic symptoms of schizophrenia. One compound known as G22355 failed to help schizophrenics but did improve depression symptoms. This drug became imipramine, sold under the brand name Tofranil. It became available in Europe in 1958 and in the United States in 1959. Imipramine was the first of a group of related drugs called tricyclic antidepressants (TCAs) because of their related chemical structure.

By the late 1960s, researchers realized that depression seemed to be related to a deficit of neurotransmitters in the brain, especially serotonin. Neurotransmitters are chemicals secreted at the ends of neurons (nerve cells) that allow an electrical impulse traveling down the nerve to jump across a synapse (gap) to the next nerve cell until the impulse reaches its destination. Common brain neurotransmitters include serotonin, dopamine, and norepinephrine. Once the electrical impulse has jumped the gap, these chemicals are reabsorbed by the ends of the neurons. TCAs partially interfered with the proteins needed for this reabsorption, thus increasing the concentration of neurotransmitters in the synapse. After several weeks of treatment, many, but not all, people found their symptoms of depression had improved.

Selective serotonin reuptake inhibitors (SSRIs) were a further refinement in antidepressant drug therapy. Instead of slowing the uptake of multiple neurotransmitters, SSRIs used a different biochemical mechanism to selectively favor an increase in serotonin over norepinephrine. This approach was more effective and had fewer side effects in many people, although with these drugs, as with TCAs, it usually took several weeks of treatment to see any improvements. The first SSRI, fluoxetine (Prozac), was marketed in the United States in 1988.

Variations on SSRIs were developed, and this class of drugs became the mainstay of depression treatment. However, in 2019 a new drug, esketamine (Spravato), was approved by the FDA for depression. Spravato is a fast-acting ketamine-based antidepressant nasal spray. Ketamine was approved by the FDA as a surgical anesthetic in 1970, but injectable ketamine has been used off-label to treat depression since the early 2000s. The drug targets a neurotransmitter in the brain called glutamate. Glutamate allows small groups of neurons to connect, communicate, and coordinate with each other. Although the drug has some negative side effects, these tend to disappear after about an hour and improvement in depression symptoms is rapid. Spravato is not appropriate for every person with depression, and it has a moderate potential for abuse.

MDD is a common problem that can lead to suicide. In the United States, suicide is the tenth leading cause of death, and worldwide before COVID-19, it was considered the third major cause of death. Between 10% and 14% of the world's population is expected to experience at least one episode of MDD. Even when successfully treated, patients are at risk for relapses.

Antidepressants are the most common class of drug prescribed in the United States. The market is huge and profitable, and their use is steadily increasing. In 2019, 3.4 million Americans filled at least one prescription for an antidepressant drug, with West Virginia leading all states in antidepressant usage. In 2020, the worldwide market was about $15 billion, and it was expected to grow by about 35% over the next five years. Many of the people receiving prescriptions for antidepressants have had no formal psychiatric evaluation and diagnosis. In the 50 years since the accidental discovery of isoniazid, the use of psychotherapeutic drugs has changed the image and treatment of mental illness and returned large profits to the companies that make these drugs.

FURTHER INFORMATION

Brown, Walter, and Maria Rosdolsky. "The Clinical Discovery of Imipramine." *American Journal of Psychiatry* 172, no. 5 (May 2015): 426–429. https://ajp.psychiatryonline.org /doi/epdf/10.1176/appi.ajp.2015.14101336

MacElvert, Raleigh. "The Past, Present and Future of Using Ketamine to Treat Depression." *Smithsonian Magazine*, May 24, 2022. https://www.smithsonianmag.com/science-nature /a-brief-history-of-ketamines-use-to-treat-depression-180980106

Nature Publishing Group. "Depression and Its Treatment." YouTube, December 19, 2014. https://www.youtube.com/watch?v=Yy8e4sw70ow

41. First Automated Blood Counter, 1953

The first automated blood counter, now called a hematology analyzer, was invented in 1953 by Wallace Henry Coulter (1913–1998). The device called the Coulter Counter revolutionized medical technology and served as the basis for what became a multimillion dollar company.

Wallace Coulter was born in Little Rock, Arkansas, in 1913. The Coulter family was an average middle-class family. Coulter's mother taught kindergarten. His father was a train dispatcher. The only notable thing about Wallace as a child was his interest in numbers and in putting things together to build gadgets.

Coulter graduated early from high school, at age 16. He attended Westminster College in Fulton, Missouri, for one year and then transferred to Georgia Institute of Technology for two years to study electronics. He finished his college education at Hendrix College in Conway, Arkansas. After college, Coulter worked maintaining radio equipment for station WNDR in Memphis, Tennessee. He later worked for General Electric in Chicago where he serviced and repaired medical equipment. This gave him the understanding of medical laboratory procedures that he needed to create his breakthrough device.

General Electric eventually offered Coulter the opportunity to do sales and service work in Asia. He spent six months in Manila, six in Shanghai, and six more in the Philippines, after which he was transferred to Singapore. In February 1942, the Japanese bombed and eventually captured Singapore. Wallace was fortunate to get out just before the Japanese took over. Because of war in Europe, he was unable to return directly to North America. His trip home turned into an almost year-long odyssey that took him to India, Africa, and South America before he finally reached the United States.

After World War II ended in 1945, Coulter worked for Raytheon and Middleman Electronics in Chicago. At the same time, he and his brother Joseph (d. 1995), an electrical engineer, worked at night on experimental projects in their basement. Coulter's big breakthrough came in the late 1940s when he developed a theory called the Coulter Principle. He attempted to patent his idea but was turned away by many patent attorneys who did not believe the idea was patentable. He persisted, and a patent was finally granted in October 1953. Of the 85 patents he received during his lifetime, this one was the most significant.

The Coulter Principle was the breakthrough that made the Coulter Counter possible. This principle says that changes in an electric current can be used to determine the number and size of particles suspended in a liquid that transmits electricity. Cells (or other particles) conduct electricity much more poorly than the fluid in which they are suspended. If the cells are drawn through a very tiny hole or microchannel, they will cause a brief but measurable decrease in an electric current across the microchannel. The decrease will be in proportion to the cell's size. The Coulter Counter used this discovery to count blood cells for what is known as a complete blood count or CBC.

The first Coulter Counter, the Model A, consisted of two fluid-filled chambers separated by a microchannel. Blood to be counted was diluted so that cells crossed through the microchannel one at a time. A vacuum pump drew the fluid containing the blood cells slowly across the channel. An electric current ran across the microchannel. When a cell passed through the channel, the electric current decreased, and the decrease was proportional to the size of the cell. (Red blood cells and white blood cells differ both in size and shape.) By recording each change in the electric current, the number of each type of cell could be determined. In 1953, Coulter sent two prototypes of his Coulter Counter to the National Institutes of Health (NIH) to be evaluated. The NIH found that the counters were more accurate and more convenient than manual counting.

The Coulter Counter was an immediate success. Coulter and his brother incorporated in 1958 and moved their company to Florida. They continued to improve blood analyzers and also moved into the production of other types of laboratory equipment. The company remained family owned until 1997 when it was sold to Beckman Instruments, which today is known as Beckman Coulter.

Wallace Coulter never married. When the company was sold, he set aside money to pay all his employees worldwide $1,000 for each year they had worked for the Coulter Corporation. Much of the remaining money from the sale went toward establishing the Wallace H. Coulter Foundation whose mission is to improve health care through research and engineering.

HISTORICAL CONTEXT AND IMPACT

The invention of the microscope (see entry 7) allowed Antonie van Leeuwenhoek (1632–1723) to examine blood. He saw many red globules (red blood cells). He could also see that blood contained other cells, but the resolution of his lenses was not good enough to count them. The first clinical blood count is credited to Karl von Vierordt (1818–1884), a professor of medicine at the University of Tübingen in Germany. He published his work in 1852. Another advance occurred in the late 1870s when German scientist Paul Ehrlich (1854–1915) devised a method of dyeing blood cells that allowed him to distinguish between multiple types of white blood cells. He noted that the quantity of different kinds of blood cells changed with different diseases. For the first time, counting different types of blood cells became a diagnostic tool.

For years before the Coulter Counter was invented, laboratory technicians spent hours hunched over microscopes counting blood cells smeared on a special slide with a grid to aid the counting process. Counting a single sample took at least half an hour. Accuracy depended on the counter's training, skill level, and ability to concentrate uninterrupted for long periods. Various studies have found that at least 10% of manual counts are significantly inaccurate. When the first commercial Coulter Counter became available in 1953, it could do the same counting job in 10 minutes and with greater accuracy. It also freed laboratory technicians to do other jobs.

Almost immediately, technical improvements began to be made to the original Coulter Counter, but the basic principle for counting cells remained the same. Gradually, devices to measure hemoglobin, the oxygen-carrying molecule in the blood, were incorporated into blood analyzers so that more information could be acquired from the sample. The Coulter Counter was also modified for industrial use so that it could measure particles in other fluids such as paint, beer, and chocolate.

A big change in blood analyzers came around 1970, when new techniques using a laser beam instead of an electric current were developed to count and differentiate cells. This new process was called flow cytometry, and it could count thousands of cells per second. In addition, fluorescent tags and dyes could be added to cells to identify, count, and separate different types of white blood cells. The drawback to this method is the amount of time it takes to prepare the blood sample.

Through a combination of the Coulter Principle and flow cytometry, engineers have developed a machine capable of identifying and counting five different white blood cell types. These machines are used in many commercial laboratories. More advanced machines exist that can detect seven or more types of white blood cells and cell subtypes. The most advanced machines are extremely expensive and currently are used primarily in research environments. Researchers are excited about the possibility of counting cell subtypes as a way of detecting and evaluating cancer and the effectiveness of disease treatments.

Wallace Coulter started a revolution in blood counting. Today the complete blood count remains one of the most commonly ordered and useful laboratory tests. The basic test is quick and inexpensive. The results give a general indication of the patient's health and help physicians decide when additional tests are needed and if so, which ones.

FURTHER INFORMATION

Green, Ralph, and Sebastian Wachsmann-Hogiu. "Development, History, and Future of Automated Cell Counters." *Clinics in Laboratory Medicine* 35, no. 1 (March 2015): 1–10. https://www.researchgate.net/publication/272186770_Development_History _and_Future_of_Automated_Cell_Counters

Robinson, J. Paul. "Wallace H. Coulter: The Man, the Engineer, and the Entrepreneur." YouTube, November 5, 2013. https://www.youtube.com/watch?v=kjVra3GTxtQ

Sullivan, Ellen. "Hematology Analyzer: From Workhorse to Thoroughbred." *Laboratory Medicine* 37, no. 5 (May 2006): 273–278. https://academic.oup.com/labmed /article/37/5/273/2504472

Wallace H. Coulter Foundation. "Biography." 2022. https://whcf.org/wallace-h-coulter /wallace-biography

42. First Successful Human Organ Transplant, 1954

The first successful human organ transplant occurred on December 23, 1954, when Joseph E. Murray (1919–2012) and his colleagues John Hartwell Harrison (1909–1984) and John P. Merrill (1917–1984) performed a kidney transplant at Peter Bent Brigham Hospital (now Brigham and Women's Hospital) in Boston, Massachusetts. Murray, the team leader, had developed the surgical techniques for kidney transplantation that made this operation possible.

Joseph Murray was born in Milford, Massachusetts. He decided at a young age to become a surgeon, although he chose a liberal arts curriculum at his undergraduate college. After graduating from Harvard Medical School and completing one year of internship in plastic surgery, he was drafted into the Army Medical Corps to serve in World War II.

Murray was posted to Valley Forge General Hospital in Pennsylvania. Here he worked under chief surgeon James Barrett Brown (1899–1971), the first person to successfully perform a permanent skin graft between identical twins. At Valley Forge, Murray treated Charles Woods (1920–2004), a military pilot. Woods was delivering aviation fuel to China when his plane crashed on takeoff. Over 70% of his body was burned. Because of the extensive burns, Murray's only choice was to graft cadaver skin onto Woods.

Normally the body's immune system rejects foreign skin in about 10 days, but to the surprise of both Murray and Brown, Woods's body tolerated the foreign skin grafts long enough to grow new skin of its own. This sparked Murray's interest in transplant surgery. After 24 reconstructive surgeries, Woods left the hospital, went on to build a financial empire, and ran unsuccessfully for governor of Alabama and several national offices.

After the war, Murray returned to Boston to continue training in reconstructive surgery but found that he was increasingly drawn to transplant surgery, especially as a solution to end stage renal (kidney) disease. In the days before dialysis or kidney transplantation, end stage renal disease was always fatal. Murray wondered whether a transplanted kidney, if accepted by the body, would function properly. By performing kidney transplants on dogs, whose immune system is somewhat tolerant of foreign tissue and less likely to reject transplanted organs, he determined that a transplanted kidney, when correctly attached to blood vessels and the ureter, could function well.

In 1954, Murray was introduced to identical 23-year-old twins Richard Herrick (1931–1963) and Ronald Herrick (1931–2010). Richard had end stage kidney failure while Ronald had two healthy kidneys and was willing to donate one to his twin. A successful human kidney transplant had never been done before, but for the twins, it was try or let Richard die. The procedure, however, created an ethical dilemma. Removing a functioning kidney from a healthy person might cause damage to the donor. This would violate the oath doctors take to do no harm.

After consulting medical and religious leaders, Murray and the team concluded that under such life-or-death circumstances, it was acceptable to proceed. The operation was successful. Richard Herrick recovered, married his nurse, fathered two children, lived another eight years, and died from heart failure. Ronald Herrick lived to age 79 with no complications from his kidney donation.

Murray continued to work as a transplant surgeon until 1971, when he left transplant surgery to work in reconstructive plastic surgery for children. Despite this switch, in 1990, he was awarded the Nobel Prize in Physiology or Medicine based on his role in advancing organ transplantation. Murray shared the prize with E. Donnall Thomas (1920–2012), who had made breakthroughs in bone marrow transplantation to treat leukemias. As of 2022, Murray was the only plastic surgeon to ever win a Nobel Prize. In retirement, he wrote a book about his medical career. Murray died at age 93 in the same hospital where he performed his historic organ transplant.

HISTORICAL CONTEXT AND IMPACT

Stories of exchanging organs between humans, or between humans and animals, have been around for centuries in many cultures. These accounts are almost certainly fictitious, but they illustrate the fascination people have with the idea of transplantation. The process of cross-species transplantation is called xenotransplantation. It is the focus of entry 45.

In the early 1900s, German surgeon Karl Thiersch (1822–1895), who studied skin cancer and wound healing, made repeated attempts to replace damaged human skin with animal skin. He concluded that animal skin would always be rejected. Therefore, he developed a successful type of autograft, or method of transferring the patient's own skin from one part of the body to another. This overcame the problem of tissue incompatibility, but it was not applicable to internal organs. However, an updated version of Thiersch's skin-grafting technique is still used today.

In 1900, Nobel Prize winner Paul Ehrlich (1854–1915) discovered that red blood cells have proteins, called antigens, on their surface. When blood with incompatible antigens is introduced into the body, the body produces antibodies that cause the blood to clump, and the patient dies (see entry 12). As understanding of the immune system progressed, it became clear that the principle of antigen incompatibility explained the rejection of animal tissue in humans as well as the rejection of tissue from genetically unrelated humans. At the same time, another Nobel Prize winner, French vascular surgeon Alexis Carrel (1873–1944), operated on dogs to developed surgical procedures to successfully connect transplanted internal organs to the vascular system.

Only a tiny number of people needing an organ transplant have an identical twin, so researchers set out to find ways of suppressing the immune system so that the body would accept tissue from less closely related people. Total body irradiation was the first method tried. The first successful kidney transplant using this procedure occurred in Louisiana in 1959. The recipient received a kidney from his brother, lived another 25 years, and died from heart failure. However, in most people, irradiation caused bone marrow to produce abnormal blood cells that eventually were fatal. As a result, irradiation was discontinued in the 1960s.

Research on drugs to suppress the immune system produced azathioprine (Imuran), which was used in transplant patients in combination with steroid drugs in the 1960s. At the same time, techniques in tissue matching improved so that donor and recipient tissue could be more closely matched. Other better drugs followed that prevented rejection or treated infections that often developed as the result of immune suppression.

Although early transplant work focused on the kidney, control of rejection presented the opportunity to transplant other organs and expanded the pool of potential donors. A simultaneous kidney and pancreas transplant operation occurred in 1966, followed by a liver transplant the next year. In 1967, Christiaan Barnard in South Africa performed the first moderately successful heart transplant. The patient died from infection after 18 days, but the heart beat normally until his death. The first successful heart-lung transplant was performed at Stanford Medical Center in California on March 9, 1981. In Spain, a full face transplant took place in 2010, and the first trachea transplant occurred in New York City in April 2021.

Initially, hospitals found their own organ donors, but in 1984, the United Network for Organ Sharing (UNOS) was established by the U.S. Congress. It maintains a transplant patient waiting list, matches donors to patients, tracks all transplants occurring in the United States, and provides education and assistance to transplant patients, their families, and transplant professionals. As of 2019, the median survival time for kidney transplant patients was 12.4 years. For liver transplants, the median was 11.1 years, but children with liver transplants can survive for over 25 years. Adult heart transplant patients had a median survival time of 9.4 years while in children it was 12.8 years.

The success of organ transplants has resulted in an increased demand for donors. Today organs can come from cadaver donors, beating heart (brain dead) donors, and for some organs, live donors, but there are not enough donors to meet demands. At the end of 2021, more than 106,000 people remained on the waiting list for organs. To ease this situation, many states allow an opt-in system where people can indicate on their driving license their willingness to be an organ donor. Other countries have increased organ availability by making donation an opt-out system, where in every accidental death, the deceased are considered potential organ donors unless they have actively opted out of the donation program. Advances have also occurred in the preservation and transportation of donated organs. As of 2021, hearts and lungs can be preserved for up to 6 hours, livers for up to 12 hours, and kidneys for up to 36 hours.

Today successful organ transplants are, if not commonplace, not rare either. According to UNOS, in 2021 a record high of 41,354 solid organ transplants occurred in the United States, including transplants of 24,669 kidneys, 9,236 livers, and 3,817 hearts. This was achieved despite a decrease in the number of donors because of COVID-19 infections.

FURTHER INFORMATION

Adenwalla, H. S., and S. Bhattacharya. "Dr. Joseph E. Murray." *Indian Journal of Plastic Surgery* 45, no. 3 (September–December 2012): 596–597. https://www.ncbi.nlm.nih.gov/pmc/articles/PMC3580381

Brigham and Women's Hospital. "Innovations in Organ Transplantation." YouTube, August 21, 2014. https://www.youtube.com/watch?v=r9nSw6-kXBw

"Joseph E. Murray." Faculty of Medicine Harvard University, November 28, 2012. https://fa.hms.harvard.edu/files/memorialminute_murray_joseph_e.pdf

Murray, Joseph E. *Surgery of the Soul: Reflections on a Curious Career*. Canton, MA: Science History Publications, 2001.

43. First Patient Treated with Ongoing Hemodialysis, 1960

Clyde Shields (1921–1971) was the first patient in the world to be successfully treated for end stage renal disease (ESRD) by hemodialysis. ESRD is an incurable, progressive form of kidney failure. Shields, the father of three, was a 39-year-old machinist for Boeing Corporation. Aside from his ESRD, he was in good health. However, without regular hemodialysis, life-threatening wastes would build up in his body, and he would die within weeks. His historic dialysis treatment occurred at University Hospital in Seattle, Washington, on March 9, 1960. It lasted 76 hours. He continued to receive regularly scheduled dialysis for 11 years and died from a heart attack.

The role of the kidney is to remove toxins that result from cellular metabolism, eliminate excess fluids from the body, and maintain the balance of critical ions, such as sodium and potassium, in the blood. Under normal circumstances, people have two kidneys. Individuals can live healthy lives with one functioning kidney, but when both kidneys permanently fail, death soon becomes inevitable without medical intervention. That medical intervention involves hemodialysis, where a machine performs the function of the kidney by cleansing the blood and balancing body fluids several times each week for a lifetime or until a kidney transplant (see entry 42) can be performed.

Dialysis requires that blood be removed from the body and passed through a dialyzer. The machine contains a semipermeable membrane that allows toxic waste to pass out of the blood and into a special fluid called the dialysate that circulates in the machine. The dialysate is then discarded, and the cleansed blood is returned to the body.

Many of the early complications of dialysis arose from the need to access veins and arteries to remove and replace blood. Over the long term, destruction of blood vessels with each treatment and the high chance of infection made chronic (long-term) dialysis impossible. Three men—Belding H. Scribner (1921–2003), Wayne Quinton (1921–2015), and David Dillard (1923–1993)—shared their skills to solve these complications. Their breakthrough saved Clyde Shields and many others from certain death.

Belding Scribner earned a medical degree from Stanford University in 1947 and received additional training at the Mayo Clinic in Minnesota, after which he moved to the Veterans Administration Hospital in Seattle, Washington, to work with kidney patients. He persuaded the hospital to invest in a kidney dialysis machine and developed a program for treating acute (short-term) kidney failure. In 1957, he moved to Seattle's University Hospital. There he collaborated with Wayne Quinton to develop a device that would make chronic dialysis possible. This invention, now called the Quinton-Scribner shunt, created a way to access the blood of dialysis patients without having to insert a new metal cannula (thin tube) into the patient's blood vessels at each dialysis session. This substantially reduced the problems of infection and blood vessel destruction. Lifesaving chronic dialysis became a reality.

The Quinton-Scribner shunt was made possible by the development of two new products, Teflon and Silastic, a siliconized rubber. David Dillard, a cardiologist, suggested the use of these materials for dialysis because they were already used and well tolerated in cardiac patients. Blood cells would not stick to Teflon and cause blockages, and using siliconized plastic in place of rubber tubing reduced the chance of infection. Dillard also worked with Scribner on the device's design and performed the operation to install the first shunt in Clyde Shields.

The shunt consisted of one cannula with a Teflon tip inserted in an arm or leg vein and a second identical cannula inserted into a nearby artery. A short piece of Silastic tubing was attached to each cannula and was threaded out through the skin. Outside the body, these two ends were connected by a U-shaped piece of Teflon. The exterior tubing was then attached to a metal plate to stabilize it. When the patient needed dialysis, the U-shaped Teflon piece was removed and the patient was connected to the dialyzer. When dialysis was finished, the U-tube was replaced. Now there was no need to puncture a new blood vessel with each dialysis.

Scribner's idea could not have become reality without the help and creativity of Wayne Quinton, a pioneering bioengineer. Quinton's mother owned a dry cleaning business in Rigby, Idaho, and at a young age Quinton learned to repair every piece of equipment the business used. During World War II, he got on-the-job design and construction training as a draftsman working with physicists and mathematicians at Boeing. Eventually, he was hired as the sole employee of the instrument shop at Seattle's University Hospital and was responsible for keeping all the hospital's instruments in good condition. He transformed the idea of the shunt into a successful product by developing special techniques for bending Teflon tubing and working with Silastic.

Quinton went on to design and build other lifesaving medical devices, including several used in open heart surgery. He later founded a successful medical device company. Dillard continued to work as a pediatric cardiac surgeon. Scribner continued working with dialysis patients. In 1962, he established the nation's first freestanding dialysis center. Scribner lived on a houseboat on the Seattle waterfront and commuted to work by canoe. In 2003, he drowned in an apparent commuting accident.

HISTORICAL CONTEXT AND IMPACT

The basis for dialysis is the natural law that substances dissolved in a solution separated by a semipermeable membrane will, if smaller than the pores in the membrane, move from an area of greater concentration to an area of lesser concentration until the concentration on either side of the membrane is equal. This was proved by Thomas Graham (1805–1869), a Scottish chemist. In 1854, he filled a container with urine, separated it from distilled water with an ox bladder, and waited. Later, when he boiled away the water, he found that sodium chloride (salt) and urea (a waste product of protein metabolism) remained. Since these were not in the original distilled water but were in the urine, he concluded that they must have diffused across the ox bladder.

Willem Kolff (1911–2009), a Dutch physician, was studying kidney failure when World War II broke out and the Netherlands was overrun by the German army. During the war, Kolff continued his kidney studies while treating wounded soldiers. Equipment was scarce, but by 1943, he had assembled a crude dialyzer to counteract acute kidney failure. It was made of sausage skins, orange juice cans, and washing machine parts. This cobbled-together apparatus is considered the first modern dialyzer. The first 16 patients Kolff dialyzed all died, but in 1945, he dialyzed a 67-year-old woman for 11 hours. She recovered, lived another seven years, and died of causes unrelated to kidney failure.

Kolff moved to the United States in 1950 where he continued to make improvements in dialysis technology. Other inventors also made improvements in dialyzers and the dialysate solution into which toxic wastes diffused. By 1960, acute dialysis was a well-established procedure, but it took the invention of the Quinton-Scribner shunt to save people with permanent kidney failure.

Once it became clear that people with ESRD could lead healthy lives as long as they received dialysis several times each week, ethical questions arose. More people needed dialysis than could be accommodated. Who should live and who should die? Who should make the decision and on what criteria? The Seattle Hospital established what became known as the "God Committee." It consisted of six lay people and one physician who was not a kidney specialist. They evaluated potential dialysis patients on the basis of age, physical and mental health, willingness to follow a dietary and dialysis regimen, and ability to contribute toward the cost of treatment. Dialysis was expensive and was not covered by health insurance.

The first year the committee evaluated 30 patients, found 17 acceptable for dialysis, but could only accommodate 10. The other 7 died. Meanwhile, Scribner secured a grant from the Hartford Corporation to build the first non–hospital-associated dialysis clinic. It opened in January 1962 in Seattle and could serve nine patients who received overnight dialysis twice a week. In 1965, Senator Henry Jackson (1912–1983) proposed that the federal government cover the cost for dialysis patients, but his proposal went nowhere. Finally on July 1, 1973, dialysis became listed as a covered procedure by Medicare.

Throughout this time, demand for dialysis increased. Dialysis machines were modified to handle more patients and be faster. Small units made home dialysis possible. New techniques developed, including peritoneal dialysis in which the membranes of the abdomen are used to filter blood that never leaves the body and wastes are withdrawn through a peritoneal catheter. Today over 2 million people worldwide receive chronic dialysis, of which about half a million are in the United States. Many of these patients continue to hold jobs and lead relatively normal, enjoyable lives while waiting for donor kidneys to become available for transplant.

FURTHER INFORMATION

Albert and Mary Lasker Foundation. "2002 Lasker DeBakey Clinical Medical Research Award—Belding Scribner." YouTube, May 3, 2016. https://www.youtube.com/watch?v=P4yt4xDB3j0

Blagg, Christopher R. "The Early History of Dialysis for Chronic Renal Failure in the United States: A View from Seattle." *World Kidney Forum* 49, no. 3 (March 1, 2007): 482–496. https://www.ajkd.org/article/S0272-6386(07)00116-3/fulltext

Fresenius Kidney Care. "What Is Dialysis?" 2022. https://www.freseniuskidneycare.com/treatment/dialysis

NWKidney. "A Tour of Northwest Kidney Centers' Dialysis Museum." YouTube, September 9, 2020. https://www.youtube.com/watch?v=xgi8E7GjgbY

44. First Oral Contraceptive, 1960

The first oral contraceptive, commonly called "the pill," was manufactured by G. D. Searle and Company and sold under the name Enovid after it received approval from the Food and Drug Administration on June 23, 1960. The birth control pill came about through the work of four unusual people: Margaret Higgins Sanger (1879–1966), Katharine Dexter McCormick (1875–1967), Gregory Pincus (1903–1967), and John Charles Rock (1890–1984). The unlikely circumstances that brought these four people together resulted in a medical first that changed the lives of millions of women and families.

Margaret Sanger (also known as Margaret Sanger Slee after divorce and remarriage) was a feminist, birth control activist, sex educator, and nurse. Her interest in birth control, a term she coined, was driven by two events. She blamed her mother's death at age 49 on 18 pregnancies and 11 live births, and she saw the desperation of poor women experiencing unwanted pregnancies or botched abortions in her work as a nurse. Sanger was an antiabortionist, but she believed all women should have control over their own bodies and that the best way to prevent abortions was to prevent unwanted pregnancies.

In 1914, Sanger published a magazine called *The Woman Rebel*, which contained articles on how to prevent pregnancy. This information was considered "lewd," "obscene," and "immoral" under the 1873 Comstock Act. Sanger was arrested but not convicted. Undeterred, in 1916, she opened the first birth control clinic in the United States in Brooklyn, New York. Under state and federal laws, the clinic was illegal. Sanger was arrested twice for distributing contraceptives and running a public nuisance. The second time, she served a 30-day workhouse sentence.

Sanger founded the American Birth Control League, which became Planned Parenthood Federation in 1921. Two years later, the first legal birth control clinic

Margaret Sanger's clinic in Brooklyn where Sanger was arrested for distributing birth control information. Her campaign for birth control resulted in the first oral contraceptive. (Library of Congress)

opened in the United States using a loophole that allowed physicians to legally prescribe contraceptives "for medical reasons." At the time, relatively ineffective condoms (see entry 4) were the main form of contraception available in the United States. Sanger's dream of a contraceptive pill that would put reproduction completely under a woman's control seemed far away. Then she met Katharine McCormick, the second woman who played an essential role in the development of the pill.

Katharine McCormick was an educated (BSc in biology from MIT) upper-class woman who gave up her dream of becoming a doctor to marry Stanley McCormick in 1904. Stanley was heir to the International Harvester fortune worth about $20 million ($600 million in 2022). Stanley's mental health began to deteriorate almost immediately after the marriage. By 1906, he was declared legally incompetent and institutionalized. This left Katharine with access to part of Stanley's fortune, which she used to further her interests in women's issues. On his death in 1947, she gained full control of his estate.

Katharine McCormick campaigned for women's rights and gender equality. She was vice president and treasurer of the National American Woman Suffrage Association and worked for the ratification of the Nineteenth Amendment to give women the right to vote. In 1917, McCormick and Sanger met through their shared interest in gender equality and birth control.

Europe was more advanced in legalizing contraceptive methods than the United States. Whenever McCormick went to Europe, which she did frequently, she would place a large order for diaphragms to be shipped to her family home in Switzerland. There her seamstress would sew them into the lining of McCormick's coats so that she could smuggle them into the United States. In the 1920s, Katharine McCormick illegally imported over 1,000 diaphragms.

Sanger shared the dream of a contraceptive pill with McCormick. McCormick had the money to fund the project, but neither woman had the science background to do the necessary research. That fell to researcher Gregory Pincus and physician John Rock.

Gregory Pincus (1903–1967), the son of Jewish Russian immigrants, was a brilliant scientist and an authority on mammalian reproductive hormones. In 1936, when he was an assistant professor at Harvard, he published the results of an experiment on in vitro fertilization. He removed an egg from a rabbit, exposed it to a form of estrogen, and when it began to develop without being fertilized by a sperm, he replaced the egg in the mother rabbit who later gave birth. The results of this experiment were reported in the media in a way that suggested that Pincus would soon be experimenting with creating human life in a test tube. The resulting negative publicity destroyed his chance of getting tenure at Harvard.

Pincus moved to Clark University in Worcester, Massachusetts. Later, he and Howard Hoagland (1899–1982), a distinguished neuroscientist, founded the Worcester Foundation for Experimental Biology. Pincus was joined by researcher Min Chueh Chang (1908–1991), and they continued their research on female reproductive hormones.

Margaret Sanger met Gregory Pincus at a dinner in 1951. Pincus was aware of studies that showed progesterone could disrupt ovulation. Sanger, impressed by Pincus's expertise, convinced Planned Parenthood to give him a small grant to explore these findings. He and Chang tested over 200 materials before they found a progesterone-related substance in the root of wild Mexican yam (*Dioscorea villosa*) that could consistently stop ovulation in laboratory animals. Planned Parenthood had little interest in pursuing this research, so in 1953, Sanger arranged for McCormick to meet Pincus. After the meeting, McCormick gave Pincus yearly grants equivalent to about $1.5 million in 2022 dollars.

To make a practical birth control pill, it was essential to find a laboratory-made chemical that would mimic the effect of the wild yam root. Pincus eventually found synthetic hormones—manufactured by pharmaceutical companies for a different purpose—that could prevent ovulation in women, but he needed a physician to help with human research. John Rock was his unlikely choice.

Rock was a devout Catholic obstetrician-gynecologist at a time when the Catholic Church forbade all types of birth control. Rock had established the Fertility and Endocrine Clinic at the Free Hospital for Women in Brookline, Massachusetts. His expertise in reproductive hormones came from helping infertile women become pregnant and not from trying to prevent pregnancy. Nevertheless, during his medical residencies, Rock had seen many women with unwanted

pregnancies. Despite his faith, he believed that practicing birth control was better than bearing unwanted children, so he accepted Pincus's offer to join the
research team. Contraception research was illegal under Massachusetts state law,
so they called their project "fertility research."

In 1956, the Pincus team submitted an application to the Food and Drug
Administration (FDA) for approval of the first birth control pill, only they
did not call it a contraceptive. They asked that it be approved for "menstrual
irregularities," and in 1957 it received approval for that purpose. With additional research to reduce side effects and clinical trials done in Puerto Rico
and Haiti because contraceptive research was still illegal in the United States,
the first birth control pill, marketed under the name Enovid, was approved in
1960 by the FDA. Within two years, 1.2 million American women were taking the pill.

HISTORICAL CONTEXT AND IMPACT

Contraception was not legally restricted in the United States until the
1870s, when a middle-class purity movement encouraged laws restricting contraception on the grounds that it encouraged prostitution, promiscuity, and the
spread of venereal diseases. Anthony Comstock (1844–1915), founder of the
New York Society for the Suppression of Vice, an organization to police public
morals, succeeded in 1873 in getting the federal government to pass laws that
made it illegal to use postal mail or any transportation system to send material
that was "obscene, lewd, or lascivious." This included everything from human
anatomy books to birth control devices. Laws also prohibited the printing and
publication of any materials related to abortion, birth control, and venereal
disease.

Comstock got himself appointed a U. S. Postal Inspector and proceeded to vigorously enforce the law. Individual states also passed laws limiting the transportation, selling, and advertising of contraceptives. This remained the case until 1965
when the U.S. Supreme Court in *Griswold v. Connecticut* declared that these laws
violated the right to marital privacy. However, until the 1970s, it remained illegal in some states to prescribe contraceptives to unmarried women.

Just as the laws changed, so did the birth control pill. Enovid contained large
doses of progesterone and estrogen equivalents. Many women had side effects,
some serious, including an increased risk of blood clots (deep vein thrombo-
sis). Gradually the combination of estrogens and progestins was modified and
reduced, and side effects decreased. A progestin-only pill, sometimes called a
mini-pill has also been developed to eliminate all side effects caused by estrogen,
although to be effective the progestin pill must be taken on a strict timetable.
Over time, different pill administration schemes developed, so that today there
are 21-, 28-, 90-, and 365-day pill regimens. In addition, contraceptive hormones
can be administered by means of skin patches, vaginal rings, injections, and long-
term implants. In 2019, 25–30% of women between the ages of 15 and 49 years in

the United States used oral contraceptives, and many others used alternate forms of hormonal birth control.

FURTHER INFORMATION

Dhont, Marc. "History of Oral Contraception." *European Journal of Conception & Reproductive Health Care* 15, no. S2 (November 2010): S12–S18. https://www.tandfonline .com/doi/full/10.3109/13625187.2010.513071

"History of the Female Birth Control Pill." Sexucation. YouTube, October 26, 2017. https://www.youtube.com/watch?v=W9Bc7De9s04

Horwitz, Rainey. "The Woman Rebel (1914)." *Embryo Project Encyclopedia*, May 16, 2018. https://embryo.asu.edu/handle/10776/13063

45. First Successful Human Xenotransplant, 1963

Xenotransplantation is the insertion of tissue from one species into a different species with the goal of replacing a damaged or diseased organ with a healthy one. The first successful xenotransplant was the implantation of a chimpanzee's kidney into a 23-year-old woman with renal failure. The surgery was performed in 1963 at Tulane University in New Orleans, Louisiana, by Keith Reemtsma (1925–2000).

A functioning kidney is essential for life. The kidneys remove toxins and waste products from blood and help to regulate chemical and fluid balance in the body through the excretion of electrolytes (mineral ions) in urine. A person can live with one functioning kidney, but failure of both kidneys is, without intervention, fatal.

The first successful human-to-human kidney transplant occurred between identical twins in 1954 (see entry 42). This operation was lifesaving, but donor kidneys remained scarce. Ongoing hemodialysis, or blood cleansing (see entry 43), that keeps many people alive today as they wait for a donor kidney had not yet been established. To Keith Reemtsma, the logical solution was to substitute a healthy animal kidney for a failed human one.

Reemtsma, the son of a Presbyterian minister and missionary, was born in California but spent part of his youth on the Navajo reservation in Arizona, where he was educated in a one-room school that had classes only up to eighth grade. Later he attended boarding school, graduated from Idaho State University, and earned a medical degree at the University of Pennsylvania where he trained as a pediatric surgeon under C. Everett Koop (1916–2013), the doctor who later became the 13th Surgeon General of the United States. Reemtsma's medical career was interrupted by the Korean War, where he served as a Navy doctor with the Marine Corps. Later in life, he and his wife claimed that he was the model for Dr. Hawkeye Pierce in the television series M*A*S*H*.

In the early 1960s, most surgery involved either removal of abnormal tissue or amputation of a body part. As a surgeon, Reemtsma was more interested in repairing or replacing damaged blood vessels. Over time, this interest extended to replacing failing organs. He chose chimpanzees as organ donors because of their size and evolutionary closeness to humans.

From November 1963 through February 1964, Reemtsma transplanted chimpanzee kidneys into six patients with renal failure who were close to death. Four of the patients lived for only a short while. One man survived for nine weeks, but a 23-year-old woman survived and was healthy enough to return to her job as a teacher, even as she continued to take immunosuppressant drugs that put her at risk for acquiring other infections. Nine months after her xenotransplant, the woman suddenly died. On autopsy, her body showed no signs of having rejected the chimpanzee kidney. The cause of death has been ascribed by various sources to infection or to an electrolyte imbalance. An electrolyte imbalance could have occurred if the kidney was producing urine, but the quantities of ions such as sodium and potassium that were removed in the urine were out of balance.

A few years later Reemtsma became chairman of the department of surgery at the University of Utah, where he worked on developing an artificial heart. He then spent 23 years as chairman of the Department of Surgery at Columbia University College of Physicians and Surgeons, where he established a transplant program. Keith Reemtsma died in 2000 of liver cancer. He is credited with being one of the people who helped to change surgery from a slice, dice, and excise mentality to a repair and replace approach.

HISTORICAL CONTEXT AND IMPACT

From early times, people have imagined the joining of human and animal bodies—Anubis the jackal-headed Egyptian god, Alkonost, the woman-headed bird of Slavic folktales, Greek centaurs and satyrs—but real-life experiments in exchanging material between animals and humans came much later. Frenchman Jean-Baptiste Denys (1643–1704) transfused calf blood into a mentally ill man, who died after the third transfusion (see entry 12). Denys was charged with murder and later acquitted, but this experience ended his human-animal experiments and discouraged others from attempting similar research.

In the 1800s, surgeons tried grafting skin from various animal species, including dogs and birds, to cover human wounds. Some of these grafts "succeeded" in that they protected the wound long enough for the body to grow new human skin, but none of the grafted material became a permanent part of the body— probably a good thing for the patients as it eliminated the need to explain fur or feathers sprouting from a hand or leg.

The real breakthrough in organ transplantation came when French surgeon Alexis Carrel (1873–1944) developed a method of stitching blood vessels together. Not only did this save the lives of the wounded but for the first time a transplanted organ could be integrated into the circulatory system and have the chance of functioning normally. Carrel won the Nobel Prize in Physiology or Medicine in 1912 for the development of this repair technique. But connecting a foreign organ to the circulatory system was not enough to yield success. It took a better understanding of how and why the immune system identified and rejected foreign tissue and the development of immunosuppressant or antirejection drugs to make even same-species transplantation possible.

Reemtsma's success with the chimpanzee-human xenotransplant opened up a new set of possibilities, but unfortunately, it turned out to be an outlier event. Thomas Starzl (1926–2017) tried transplanting baboon kidneys into human patients. All the patients quickly died. Later he tried transplanting baboon livers into humans without any long-term success, although he went on to have an outstanding career in human-to-human liver transplantation and in advancing the development of immunosuppressant drugs.

In 1964, James Hardy (1918–2003) at the University of Mississippi performed the first heart xenotransplantation, using a chimpanzee's heart. The patient was semicomatose even before the operation and died in the operating room. This operation definitively established that a chimpanzee heart was too small to support the human circulatory system, but it also raised serious questions about medical ethics and informed consent.

After Hardy's failure, interest in heart xenotransplantation dropped. However, South African surgeon Christiaan Barnard (1922–2001) did succeed in a human-to-human heart transplant in 1967. His patient lived for 18 days. In a second heart transplant the following year, the patient was well enough to go home from the hospital and lived another 1.5 years.

In 1984, xenotransplantation became a topic of controversy when Leonard Baily (1942–2019) transplanted a baboon heart into an infant known as "Baby Fae" who had been born with a fatal heart defect. The baby lived only 20 days. The operation aroused anger among animal rights activists and triggered hard questions from ethicists.

After failed attempts to transplant primate hearts into humans, pigs became the donors of choice. Pigs are abundant and cheap. Raising one baboon and transplanting its organs would cost an estimated $500,000 in 2021. The smaller breeds of pigs are easy to care for, at least when compared to chimpanzees or baboons, and their organs are about the right size for humans. By 1995, pigs could be genetically modified so that a protein made in their cells that reacts with human blood was eliminated.

In 1997, just when it looked like a breakthrough in xenotransplantation was about to occur, researchers discovered that every pig cell contained a virus called porcine endogenous retrovirus (PERV) that had the potential to infect humans. Based on fear that the virus could start an epidemic, the U.S. Food and Drug

Administration (FDA) temporarily halted all xenotransplants, and in 2000, a worldwide xenotransplant ban went into effect. This ban has since been lifted or modified, or requires a special exemption permit in many countries.

Despite the ban, researchers continued to develop genetically modified transgenic pigs, also called "knockout pigs" because genetic modification "knocks out" the production of certain proteins that negatively react with the human immune system. These clinical-grade pigs are raised on special farms in hygienic environments. In 2022, the only farm in the United States approved to raise clinical-grade pigs was in Blacksburg, Virginia.

Using new gene editing tools, researchers were able to create and clone a modified pig in which four pig genes were knocked out and six human genes were knocked in or added to the pig genome. Kidneys from these pigs were transplanted into several brain-dead humans who were maintained on ventilators for a few days. The kidneys functioned well and were not immediately rejected, nor were any pig viruses found in the human recipients during the short time the kidneys were implanted in them.

Finally on January 7, 2022, surgeons at the University of Maryland Medical Center, with special permission from the FDA, transplanted a genetically modified pig heart into David Bennett, aged 57, after he was determined to be ineligible for a human heart transplant. Bennett survived for two months, giving transplant surgeons hope that in the future pigs could provide desperately needed organs. After his death, however, it was discovered that Bennet had been infected with a pig cytomegalovirus (not PERV). Since Bennett was already severely ill, it is unclear what role the virus played in his death. However, transplant surgeons considered it a positive sign that Bennet's immune system showed no signs of rejecting the pig heart.

Far more people need organ transplants than there are donors. In 2021, a total of 39,717 organ transplants were performed in the United States, including 28,401 kidney transplants and 3,800 heart transplants. But more than 106,200 people remained in need of donors, with an average of 17 dying each day because donated organs were not available. Some, because of the long time they had spent on the wait list, had become so ill that they could no longer be considered good transplant candidates. Researchers hope that in the future genetically modified pigs will be able to fill the organ donation gap and extend the life of people with organ failure.

FURTHER INFORMATION

Cooper, David K. "A Brief History of Clinical Xenotransplantation." *International Journal of Surgery* 23 (November 2015): 205–210. https://www.ncbi.nlm.nih.gov/pmc/articles/PMC4684730

Demystifying Medicine Macmaster. "The Science of Xenotransplantation and How It Can Change Our World." YouTube, March 30, 2022. https://www.youtube.com/watch?v=Age2eGLZ8nQ

PBS. "Organ Farm." *Frontline*, March 27, 2001. Transcript. https://www.pbs.org/wgbh
/pages/frontline/shows/organfarm/etc/script1.html
Whiteboard Doctor. "Pig Heart Successfully Transplanted into Human Male—
Breakthrough in Xenotransplantation." YouTube, January 13, 2022. https://www
.youtube.com/watch?v=d6j6c_60X38

46. First Disease Cured Using Stem Cells, 1963

Leukemia was the first disease to be cured by the use of stem cell transplantation. In 1963, Georges Mathé (1922–2010), a French oncologist and immunologist, cured a patient of leukemia using a bone marrow transplant. The patient lived for 20 months and at autopsy had no sign of the disease.

Leukemias are cancers of blood cells. Blood cells are made in bone marrow found in the hollow interior of bones. Marrow makes up about 4% of body weight. Our bodies contain two types of bone marrow, red and yellow. Yellow bone marrow makes cells that become bone, cartilage, or fat and is not involved in blood production or leukemia. Red bone marrow contains cells called hematopoietic stem cells (HSCs). Stem cells are pluripotent, meaning they can differentiate and mature into many types of specialized cells. HSCs in red marrow develop into red blood cells, five major types of white blood cells, and platelets. Red blood cells deliver oxygen and remove waste from cells. White blood cells fight infection. Platelets are cell fragments that help blood clot. Healthy blood cells have limited lifetimes, so red bone marrow churns out about 300 million replacement cells each day.

Leukemias are cancers of white blood cells. In people with leukemia, the white blood cells do not mature or function properly but continue to grow and divide rapidly using nutrients needed by other cells. Because there are multiple kinds of white blood cells, there are multiple types of leukemias. Acute lymphocytic leukemia is the most common form of the disease. It occurs most often in children but sometimes develops in adults. Before George Mathé's experiments in bone marrow transplantation, a diagnosis of leukemia meant that it was just a matter of time until the individual died.

Georges Mathé did not suddenly have a brainstorm and one day start transplanting bone marrow into leukemia patients. Like most researchers, he worked from a knowledge base built up over time. Soon after X-rays were discovered in 1895 (see entry 25), scientists discovered that X-rays could kill cells. They experimented to determine if radiation would kill cancer cells. It would, but the dose needed to kill cancer cells also killed healthy cells and caused other complications, including radiation poisoning and sometimes death.

In 1957, Mathé was able to cure mice of leukemia. First, he irradiated inbred mice who had leukemia with a dose that would kill bone marrow cells. Next, he injected the mice with bone marrow from healthy mice. The new bone marrow made healthy blood cells, and because the mice were inbred, their bodies were similar enough that they did not reject the new cells. But mice are not men. The amount of radiation needed to kill bone marrow cells in humans was likely to kill them, so human experimentation was ethically impossible. Then an accident created the perfect human experiment.

In late October 1958, six young researchers at a nuclear reactor in Vinca, near Belgrade, Yugoslavia (now Serbia), were accidentally exposed to high doses of radiation. There was no known treatment. The scientists' conditions worsened, and it looked as if they would die. As a last attempt to save them, they were flown to Paris where they were treated experimentally by Georges Mathé.

The researcher exposed to the largest dose of radiation died. The other five were given bone marrow transplants from unrelated donors, a process called allogenic transplant. The donors' blood types were matched to the recipients' blood type, but that was the only available form of matching at the time.

On November 11, 1958, Radojko Maksić, a 24-year-old physics graduate student, received the first allogenic human bone marrow transplant from donor Marcel Pabion. The other researchers were also treated. Afterward, all the treated scientists developed complications, most likely related to the inability to match specific cell surface proteins as is done today. Still, they all survived, as did all the donors.

Rosanda Dangubić, one of the researchers, had a healthy baby girl a few years after her bone marrow transplant. Radojko Maksić, the first transplant recipient, was 82 years old and living in Belgrade in 2016. This nuclear accident allowed Mathé to show conclusively that hematopoietic stem cells artificially introduced into the body would take up residence in the bone marrow and make healthy new replacement cells and that bone marrow donors were not harmed.

This emergency experiment set the stage for Mathé to attempt other human allogenic transplants. In 1963, he succeeded in curing a girl of leukemia by using a bone marrow transplant from an unrelated donor. He exposed the girl to radiation and various chemicals that would wipe out her bone marrow cells and then injected her with new donor marrow cells. His proof that the new stem cells had differentiated and were producing healthy cells was absolute. The girl's blood type changed from the type she had at birth to the donor's blood type. Unfortunately, the girl died from a brain infection 20 months later, but at autopsy her body showed no signs of leukemia. This was the first disease cured by stem cell transplantation.

HISTORICAL CONTEXT AND IMPACT

It is unclear who first recognized leukemia as a distinct disease. French surgeon Alfred Velpeau (1795–1867) seems to have been the first person to describe the symptoms of leukemia in an 1827 report on a 63-year-old patient. Alfred Donné (1801–1878), a French bacteriologist who developed a technique for photographing cells under a microscope, is credited with recognizing that the yet-unnamed disease was associated with failure of white blood cells to mature. In 1845, John Hughes Bennett (1812–1875) made the first written case report that described leukemia symptoms as a blood disorder. German pathologist Rudolf Virchow (1821–1902) finally named the disease in 1854 by combining two Greek words, *leukos* and *hemia*, which translate as "white blood." By 1913, four different types of leukemia had been described, and the following year Theodor Boveri (1862–1915) suggested that leukemia was a form of cancer.

Despite progress in defining leukemia, there was no cure. Blood transfusions were tried as early as 1873, but they only temporarily prolonged the patient's life and had risks of their own. X-ray exposure offered no permanent improvement and had negative side effects. Compounds containing arsenic were used as treatment until the 1930s but were uniformly unsuccessful. Little progress was made in treating the disease from the beginning of World War I until the 1950s, when a chemotherapy drug derived from mustard gas (see entry 39) caused a longer remission period in a few patients. Bone marrow transplantation was the breakthrough that promised a cure rather than simply prolonging life.

A better understanding of the need to match human blood cell antigens, the proteins on cell surfaces that distinguish self from non-self, improved the success of organ and bone marrow transplants as did better diagnostic, surgical, and infection control techniques. In 1990, the Nobel Prize committee recognized this by awarding the Prize in Physiology or Medicine to two Americans, Joseph Murray (1919–2012) and E. Donnall Thomas (1920–2012). Murray performed the first kidney transplant between identical twins in 1954 (see entry 42). Thomas also performed the first bone marrow transplant between identical two-year-old twins in 1956, although the twin receiving the transplant died after 166 days. Many in the transplant community felt that Georges Mathé should have been included in the Nobel Prize because his work showed that transplants between unrelated individuals could be successful and, in doing so, opened the possibility of a cure to the general population.

Bone marrow transplantation has come a long way since Georges Mathé first cured a person with leukemia in 1963. In 1986, the U.S. Navy established the National Bone Marrow Transplant Registry to match volunteer donors with those needing transplants. Today the program is funded through the U.S. Health Resources and Service Administration. It has facilitated more than 100,000 transplants and cooperates with other bone marrow registries worldwide.

Most of what used to be called bone marrow transplants are now called hematopoietic cell transplants. Instead of directly taking bone marrow from a donor and giving it to a recipient, donors are now given drugs that cause HSCs to leave the bone marrow and enter the general blood circulation from which they can be collected and isolated. These cells are then injected into the recipient where they migrate to bones and begin producing healthy blood cells. Banked cord blood taken from the umbilical cord immediately after a woman gives birth is another source of HSCs. These less invasive methods have encouraged thousands of people to donate their blood to successfully cure complete strangers.

FURTHER INFORMATION

Drinjakovic, Jovana. "The Story of the First Bone Marrow Transplant." *Signals*, September 15, 2016. https://www.signalsblog.ca/the-story-of-the-first-bone-marrow-transplant

Kahnacademymedicine. "What Is Leukemia." YouTube, May 6, 2014. https://www.youtube .com/watch?v=IB3iJUuxt1c

Thomas, Xavier. "First Contributors in the History of Leukemia." *World Journal of Hematology* 2, no. 3 (August 2013): 62–70. https://www.wjgnet.com/2218-6204/full/v2/i3/62.htm

Transplantation Society. "Georges Mathé, MD—The Transplantation Society 2002 Medawar Acceptance Video." YouTube, November 7, 2009. https://www.youtube .com/watch?v=DCGRQvNc2Yw&t=345s

47. First Diagnostic Computed Tomography Scan, 1971

The first diagnostic computed tomography (CT) scan on a human was performed on October 1, 1971, at Atkinson Morley Hospital in Wimbledon, England. The scanner was used to diagnose a brain tumor in a 41-year-old woman. The concept of the scanner and physical construction of the machine originated with British engineer Godfrey Hounsfield (1919–2004). The theoretical mathematical basis for the CT scanner came independently from the work of physicist Allan Cormack (1924–1998), a South African–born naturalized American citizen. These men shared the Nobel Prize in 1979. James Ambrose (1923–2006), a radiologist at Atkinson Morley, worked closely with Hounsfield to make the first functional scanner and administered the first diagnostic scan on a patient.

Godfrey Hounsfield was raised on a small farm in a village in Nottinghamshire, England, the youngest of five siblings. By his own description, he was a loner and was allowed to follow his own interests. These centered on experimenting, sometimes dangerously, with mechanical and electrical farm equipment. During this time, he also built several electronic recording machines. Despite his natural abilities, Hounsfield was an indifferent student who left school at age 16. No one would ever have predicted that he would permanently change the practice of diagnostic medicine.

Because of an interest in airplanes, Hounsfield joined for the Royal Air Force (RAF) just as World War II began. After studying books provided by the RAF, he scored high enough on an entry test to be admitted to a radar mechanics program.

Later he became an instructor at Cranwell Radar School. After the war, one of his superior officers helped him get a grant to attend the Faraday House Electrical Engineering College in London. The school offered a combination of theoretical engineering and hands-on training that suited Hounsfield's interests perfectly.

After graduation, Hounsfield joined Electrical and Music Industries (EMI). Today EMI is thought of as a record label and music publisher, but it was originally founded in 1897 to manufacture early gramophone records and had established an active electronics laboratory that later produced the first stereophonic microphone and advanced radio and television broadcast equipment. The laboratory had been repurposed, as had many in British industries, to support the war effort. In the EMI laboratory, Hounsfield worked on guided weapons and radar devices. The lab also did early computer research.

Computers intrigued Hounsfield, and as a result, he built the first all-transistor computer made in Britain. Twenty-four of these large, bulky computers were sold before technological advances made them obsolete. When a new Hounsfield computer project failed to be commercially feasible, the lab, not wanting to lose a brilliant engineer, put him on paid leave and told him to go think about other projects that might be more successful. During this leave in 1967, he conceived of the concept for the CT scanner.

When Hounsfield worked with radar, he had been intrigued by the idea that the location and external shape of an object could be determined by sending out short pulses of electromagnetic waves and analyzing how they were reflected by an object in their path. During his leave, he began to think about whether electromagnetic radiation could also be used to see inside a closed object.

X-rays are a form of electromagnetic radiation. As they pass through the body, tissues of varying densities absorb different amounts of energy. When these energy differences activate a photographic plate, bones show up clearly but soft tissue does not. Hounsfield's idea was to get an image of something inside a container, such as the soft tissue inside the brain, by sending X-rays through it from many different angles. Each angle would cover a slice of the tissue. Once the data were collected by a receiver, a computer using a complex algorithm could compile the slices into a single image that could be used to diagnose tissue abnormalities such as brain tumors.

Hounsfield took this idea back to the EMI Central Research Laboratory. By 1967, EMI was awash with money thanks to the success of the Beatles on their major label and other popular groups recorded on their subsidiary labels. The company wanted out of the computer and medical equipment industry but reluctantly gave Hounsfield £100,000 (about £1.3 million or $1.6 million in 2022) to explore his idea. He later convinced the British Department of Health and Social Security to contribute another £600,000 (£9 million or $11 million in 2022). By October 1971, after successfully demonstrating a prototype scanner on fresh cow brains obtained from a kosher butcher, he was ready to try the scanner on a human patient.

HISTORICAL CONTEXT AND IMPACT

The CT scanner was a sensation. It changed the ability to diagnose brain abnormalities overnight. The earliest technique to "see inside" the brain involved drilling holes in the skull and injecting air. The air contrasted with brain tissue so that poor quality, often useless, X-rays could be made. This process was updated in 1919 by Walter Edward Dandy (1886–1946), a doctor considered one of the founders of neurosurgery. His process, called pneumoencephalography, required removing a large amount of cerebrospinal fluid (CSF) through a lumbar puncture in the spinal column. The fluid was then replaced with a gas. The patient had to be strapped into a chair and tipped upside down and sideways so that the gas would flow into spaces in the brain formerly occupied by CSF. X-rays were taken from several angles in the hope that they would reveal an abnormality. This process was excruciatingly painful. The patient usually vomited repeatedly during the procedure, and debilitating headaches lasted for days afterward while the body made replacement CSF. The CT scanner ended this torture.

The first CT scanner was incredibly slow and primitive compared to the ones in use today. It took 160 parallel scans each 1° apart, a process that took about 20 minutes. The information was recorded on magnetic tape and physically carried back to the EMI lab where it took 21 hours for their computer to use a complex algorithm to process the data and compile an image. Nevertheless, from this first scan, the doctor was able to correctly diagnose a brain tumor. Hounsfield immediately set about making improvements.

Once the results of the CT scanner were publicized in 1972, every hospital wanted one. The first commercial scanner, which stored its information on floppy disks instead of magnetic tape, was installed in the Mayo Clinic in Rochester, Minnesota in 1973. It cost $350,000 ($2.3 million in 2022). Today it is on display at the clinic, one of only two known surviving first models. The other is in the British Science Museum. Despite the success of the CT scanner, EMI still wanted out of the medical equipment field and left it in 1979 to concentrate on the music industry.

The first scanner could only scan the head because the patient had to remain still for long periods, but by 1973, Robert Ledley (1926–2012), an American dentist turned biomedical researcher, had developed a method for scanning the whole body. By 1979, a whole body scan took only 20 seconds, a short enough time for patients to stay still and hold their breath. By the mid-1980s more than 20 companies were manufacturing CT scanners. As computers became faster, scan time became shorter. A scan that might have taken 5 minutes in 1973 now took only 1.3 seconds. In addition, improvements were made in image quality.

Spiral CT added multiple detectors circling the body, improving images until it eventually became possible to see blood vessels with clarity. Scan speed, ease of use, and image storage also improved as computers became capable of handling large amounts of data almost instantaneously. Concern over the amount of radiation the patient was exposed to during CT led to a successful industry-wide reduction in radiation dose.

By 2020, more than 80 million CT scans were performed in the United States each year. The technology has saved lives by allowing early diagnosis of tumors and has practically eliminated the need for exploratory surgery. As for Godfrey Hounsfield, despite never receiving a formal college degree, he received many scientific honors in addition to the Nobel Prize. In 1981, he was knighted, becoming Sir Godfrey Hounsfield. He never married and devoted most of his time and energy to improvements in medical electronics until he died at age 84.

FURTHER INFORMATION

Ambrose, Euan, T. Gould, and D. Uttley. "Jamie Ambrose." *British Medical Journal* 332, no. 7547 (2006): 977. https://www.ncbi.nlm.nih.gov/pmc/articles/PMC1444818

"Computed Tomography (CT)." U.S. Department of Health & Human Services National Institute of Biomedical Imaging and Bioengineering, June 2022. https://www.nibib.nih .gov/science-education/science-topics/computed-tomography-ct

"Godfrey N. Hounsfield—Biographical." Nobel Prize Organisation, 1979. https://www .nobelprize.org/prizes/medicine/1979/hounsfield/biographical

Mayo Clinic. "History of CT." YouTube, April 15, 2020. https://www.youtube.com /watch?v=M6vsBcxHPZU

Xrayctscanner. "The Scanner Story." YouTube, January 28, 2011. https://www.youtube .com/watch?v=u_R47LDdlZM

48. First Monoclonal Antibodies, 1975

The first monoclonal antibodies (mAbs) were developed in 1975 by César Milstein (1927–2002) and Georges Köhler (1946–1995). When the body is invaded by a foreign organism such as a bacterium or if body cells (self cells) become abnormal as in cancer, the immune system is stimulated to try to destroy these cells. B lymphocytes, also called B cells, are a part of the immune system. Their function is to produce proteins called antibodies. Antibodies can bind with proteins called antigens on the surfaces of non-self cells and disable them, but antibodies are not generic or interchangeable. Each B cell makes only one kind of antibody, and each antibody binds with only one specific protein on a non-self cell surface. This immune response allows us to recover from many illnesses, but sometimes the immune system is overwhelmed, and we die. This is why mAbs can be lifesaving.

mAbs are proteins made in the laboratory and injected into the body to fight disease. What makes them special is that they arise from a single (mono) cell that has been duplicated (cloned) millions of times. The identical cloned cells produce billions of identical antibodies. Those antibodies will bind to only one unique protein on a non-self cell. This specificity means that researchers must design different mAbs to target each separate disease—such as each different type of cancer.

César Milstein, the coinventor of monoclonal antibody technology, was born in Argentina. He was the son of a Jewish immigrant father from Ukraine and a mother whose parents were also Ukrainian immigrants. Milstein's parents were dedicated to doing whatever was necessary for their three children to get good educations. Milstein, although he claimed he was an indifferent student, became interested in science and earned a PhD in biochemistry from a university in Argentina and a second one from Cambridge University in England.

Milstein's initial research interest involved enzymes, but a military coup in 1962 disrupted the laboratory in Argentina where he was employed. He moved to Cambridge University where he previously had participated in a visiting research fellowship. Once back at Cambridge, his research interest shifted to immunology.

Georges Köhler, a German working at Cambridge on a postdoctoral fellowship, shared Milstein's interests in both enzymes and immunology. Together they created the first mAbs with the goal of making them a research tool. For this achievement, they were both awarded the Nobel Prize in Physiology or Medicine in 1984. They shared the prize with Danish immunologist Niels Jerne (1911–1994) who had developed the concept of antibody specificity.

When Milstein and Köhler set about creating a source of mAbs, they had a major problem. B cells are difficult to grow in laboratory cultures and have short lifespans. The researchers had to find a way to keep these cells alive, dividing, and producing enough antibodies to harvest, purify, and use for experimental purposes.

Normal healthy cells make a limited number of divisions and then die, but Milstein and Köhler were aware of a type of myeloma cancer cell that makes an unlimited number of divisions and is essentially immortal. This ability to reproduce without limits was a quality they needed in the cells that would make mAbs. They requested some immortal myeloma cells from another researcher and found that these cells would reproduce endlessly in a laboratory culture that would also support short-lived B cells.

To achieve an adequate supply of mAbs, the researchers needed to combine the immortality of a myeloma cell with the antibody-producing capability of a B cell. To do this, they first injected a mouse with an organism that caused it to produce the desired B cells. After a few days, they collected immune cells from the mouse's spleen and isolated the antibody-producing cells they wanted. Next, they developed a technique for fusing a mouse B cell with an immortal myeloma cell. This was much more difficult than it sounds. Milstein and Köhler used chemicals to force the fusion. Today it can also be done using a virus vector. The result was a single cell that Milstein and Köhler called a hybridoma.

The hybridomas had the immortality of the myeloma cell, produced a single type of antibody like the B cell, and would grow endlessly in a special laboratory culture medium from which the antibodies could be harvested, purified, and used for experimental purposes. The first successful mAbs were grown in 1975. Nine years later, Milstein and Köhler were awarded the Nobel Prize for this achievement.

HISTORICAL CONTEXT AND IMPACT

In the 1890s, German bacteriologist Paul Erlich (1854–1915) discovered that blood serum (the clear part of blood) contains particles that could neutralize the toxin from the bacterium that causes diphtheria. He named these particles "antibodies." This discovery led to the first successful attempt to use antibodies to control a disease.

Scientists found that a horse injected with a weakened form of diphtheria toxin would make antibodies against the toxin. When blood was taken from the horse and the serum isolated, it could safely be injected into a person with diphtheria. The process did not kill the horse, and the immune system of the diphtheria patient was boosted, sometimes enough to recover. This technique lacked standardization. The quantity of antibodies in the serum varied from horse to horse, but horse serum was the only available diphtheria treatment. It was introduced in the United States in 1891, and by 1895, horse antibodies against diphtheria were being produced commercially. Although an effective vaccine replaced this treatment, horse antibody serum was still used occasionally to treat diphtheria as late as 2016.

Antibodies became better understood over the next 100 years, but few commercial attempts were made to use them to treat specific diseases. This changed with Milstein and Köhler's new hybridoma technology.

In 1975, Milstein gave a presentation about his research to the British Medical Research Council (MRC). An MRC member recognized the commercial potential of the work. He asked Milstein for a preprint copy of a journal article that would soon be published in *Nature*, a prestigious scientific journal. After reading the paper, he alerted the National Research Development Corporation (NRDC), the organization responsible for filing patents for the MRC, that the Milstein-Köhler technology should be patented. British law required that a patent application be filed *before* any public disclosure of the work. The NRDC did not see any immediate application for the technology and failed to take quick action to file a patent. The *Nature* article was published, and the opportunity to patent the process in the United Kingdom was lost.

The next year Hilary Koprowski (1916–2013) at the Wistar Institute in Philadelphia, Pennsylvania, asked for a sample of Milstein's hybridomas. This was a common type of request. Scientists had been exchanging biological specimens and research materials for several hundred years. Milstein sent Koprowski the hybridomas and asked that he acknowledge Milstein for providing the source cells in any related research publications. Milstein also asked that Koprowski get his permission before sharing the hybridomas with other researchers. This never happened. Consequently, Milstein was surprised when Koprowski, oncologist Carlo Croce (1944–), and virologist Walter Gerhard filed for and were granted patents in 1979 and 1980 for making mAbs against viruses and tumors using the hybridomas Milstein had provided.

This disclosure caused an uproar in Britain because some of Milstein's work had been funded by the government and also because the NRDC had failed to file a timely patent on the work. Milstein claimed that he was annoyed but not distressed that the NRDC had not filed the patent. At that time, he would have received no money from a patent since his work was government-funded. Fury at the NRDC's failure, however, reached all the way to Prime Minister Margaret Thatcher.

After 1980, commercialization of mAbs took off. In 1986, the first monoclonal antibody drug approved by the U.S. Food and Drug Administration (FDA) was anti-CD3 muromonab (Orthoclone OKT3). It was used to prevent organ transplant rejection. In 1997, rituximab (Rituxan) became the first monoclonal antibody drug approved to treat a cancer. It was initially approved for treatment of non-Hodgkin's lymphoma. Later the approval was expanded to include several other cancers as well as rheumatoid arthritis.

By 2021, more than 100 monoclonal antibody drugs had received FDA approval. These drugs, which are expensive and time-consuming to make, are now a billion-dollar business, and new monoclonal antibody drugs are entering the market regularly. In February 2022, the FDA gave emergency use authorization to bebtelovimab, a monoclonal antibody to treat COVID-19 in patients aged 12 and older who are at risk for progression to severe COVID-19.

The early monoclonal antibody drugs had some serious negative side effects. These have been reduced by using transgenic mice as the source of B cells. Transgenic animals are animals in which a human gene has replaced an animal gene through gene editing. This reduces the chance that the human immune system will reject the drug or tissue transplant.

Today mAbs are designed to attack cells in a variety of ways. They can prevent cells from reproducing, attach toxins to cells that will then kill them, or carry a marker chemical to locate where a specific type of cell is growing in the body. mAbs are also used in blood typing and to locate blood clots in the body.

One of the most familiar uses of mAbs is in home pregnancy tests. The hormone human chorionic gonadotropin (hCG) is made as soon as a fertilized egg implants in the uterus. This hormone is excreted in urine, and, within about 10 days after conception, is often present in large enough quantities to be identified. If hCG is present in a woman's urine, it will react with mAbs on the pregnancy test stick causing a color change to indicate that the woman is pregnant.

FURTHER INFORMATION

Cleveland Clinic."Monoclonal Antibodies." November 16, 2021. https://my.clevelandclinic
.org/health/treatments/22246-monoclonal-antibodies

FuseSchool. "Monoclonal Antibodies." YouTube, July 1, 2018. https://www.youtube.com
/watch?v=M3zllm8QbCM

Marks, Lara. "The Story of César Milstein and Monoclonal Antibodies." Undated.
https://www.whatisbiotechnology.org/index.php/exhibitions/milstein/introduction
/Introduction-to-the-story-of-César-Milstein-and-mAbs

National Cancer Institute. "How Monoclonal Antibodies Treat Cancer." YouTube, June
30, 2020. https://www.youtube.com/watch?v=dxnjAc-rqz8

National Museum of American History. "The Antibody Initiative—Monoclonal Antibodies." Undated. https://americanhistory.si.edu/collections/object-groups/antibody
-initiative/monoclonal

49. First Sequencing of a DNA-Based Genome, 1977

The genome of bacteriophage phiX174 was sequenced by Frederick Sanger (1918–2013) and his team at the Molecular Research Council Laboratory of Molecular Biology in Cambridge, England, in 1977. A bacteriophage is a virus that attacks bacteria. PhiX174 attacks *Escherichia coli*. This was the first DNA-based genome to be completely sequenced.

DNA is composed of four bases (also called nucleotides): adenine (A), thymine (T), cytosine (C), and guanine (G). A always bonds with T, and C always bonds with G. An ordered combination of these base pairs form genes, and multiple genes form chromosomes. PhiX174's genetic material consists of single-stranded DNA that forms one circular chromosome made up of 11 genes composed of 5,386 bases. Sanger was able to determine the order in which bases are arranged, a process called sequencing. To do this, he had to develop a method to identify the different bases and then devise a way to order them correctly.

Frederick Sanger was born in Redcombe, England, the son of a Quaker doctor who had served as a medical missionary. In his early years, he was educated at home. Later he attended a Quaker boarding school. On entering Cambridge University, he was expected to follow his father and become a physician, but laboratory research appealed to him more than dealing with patients. Because of his sincere Quaker beliefs, he was excused from military service as a conscientious objector during World War II and remained at Cambridge where he earned a PhD in 1943.

After receiving his PhD, Sanger's next research project was to determine the amino acid sequence in insulin. Insulin is a protein. Proteins are made from only

22 different amino acids that combine in varying quantities and arrangements to form molecules. At that time, there was no method to determine the order in which the amino acids in specific proteins were connected. Insulin was chosen for analysis because it could easily be obtained in pure form from a local pharmacy. This turned out to be a lucky choice because insulin contains only a total of 51 amino acids, few enough that sequencing them seemed possible. Still, it took about 10 years to correctly determine their order.

In his work with insulin, Sanger developed new techniques for separating and identifying the amino acids. He then reassembled them in the correct order using what he called the "jigsaw puzzle method," which was based on his knowledge of the types of chemical bonds various elements form and on trial and error. For his work in clarifying the struc-

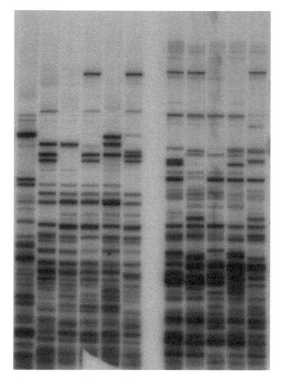

Composite gel electrophoresis of DNA showing the migration of different bases through the gel. An early DNA fingerprint showing DNA patterns from a mother (lanes 2 and 8) and her four children (adjacent lanes to the right). Lane 1 is an unrelated person. The two sets of lanes show the fingerprints revealed with two different probes that detect different types of repeated sequences. (Alec Jeffreys. Attribution 4.0 International (CC BY 4.0). Courtesy Wellcome Institute.)

ture of proteins, he won the Nobel Prize in Chemistry in 1958.

In 1962, Sanger moved to a laboratory where some of the top scientists in Britain were working on DNA-related questions. His goal was to sequence the bases in the DNA of bacteriophage phiX174. He quickly realized that there were too many bases (more than 5,000) to use the jigsaw-puzzle-trial-and-error approach.

The new methods Sanger developed for ordering DNA bases were both complex and tedious. He found a way to separate phiX174's single strand of DNA into fragments. The fragments were combined in a test tube with a radioactively labeled base and a primer, an enzyme that would catalyze new DNA synthesis. One base type was labeled in each test tube. This mixture created pieces of new

DNA complementary to the fragments Sanger had started with, and in each tube, one base—A, T, G, or C—was radioactively labeled.

Sanger used polyacrylamide gel electrophoresis to separate the labeled pieces. In this process, the fragments migrate through a sheet of gel with the smallest fragments moving the farthest. At the completion of the process, the fragments will be ordered by size on the gel plate. Next, Sanger made a radiogram or "radiation picture" of each gel plate. He then manually analyzed and combined the gel plate data. This process had to be repeated many times because he could only work with about 80 bases at a time. After 10 years, he was able to compile his findings and determine the sequence of DNA bases in the single phiX174 chromosome. This technique came to be called the Sanger Method. It was used to sequence DNA into the 1980s and serves as the basis for automated DNA analysis today.

Sanger was rewarded with a second Nobel Prize in Chemistry in 1980 for the Sanger Method of DNA analysis. He shared the prize with Paul Berg (1926–) and Walter Gilbert (1932–), both of whom were working on other aspects of DNA. As of 2022, Sanger is one of only four people to have won two Nobel Prizes. He noted that winning the second was easier than the first because after he won the first, organizations were eager to fund his research where earlier he often struggled to acquire funding.

Sanger received many other prestigious science and medical awards and was offered a knighthood, which he declined. He was known to describe himself as "just a guy who messed around in the laboratory" and stated that he had no wish to be knighted and called "Sir." Sanger retired in 1983 and died in his sleep in 2013.

HISTORICAL CONTEXT AND IMPACT

Before Sanger's success, scientists had little hope of being able to sequence the genome of organisms with a large number of bases because obtaining the information was tedious and time-intensive and interpreting the results was difficult. In 1972, Belgian molecular biologist Walter Fiers (1931–2019) decoded the base sequence for the single gene of virus MS2. This virus uses RNA as its genetic material instead of DNA. Then in 1976, the Fiers team decoded the entire RNA sequence of MS2. It has one of the smallest genomes known and contains only 3,569 bases.

Sanger's sequencing method showed that a slightly larger DNA genome could be deciphered, but the process was still unlikely to succeed when greater numbers of bases were involved. The next decoded organism, *Haemophilus influenzae*, was not sequenced until 1995. It had 1.8 million bases and was decoded only because an audacious project had been proposed in 1988. This project, called the Human Genome Project (HGP), aimed to sequence the entire human genome by 2005 along with sequencing other organisms important to biomedical research such as the mouse, fruit fly, and baker's yeast. The HGP was the molecular biology equivalent of putting a man on the moon.

The goals of the HPG were outlined in 1988. The project began in 1990 as a joint effort between the U.S. Department of Energy and the National Institutes of Health. It was expected to take 15 years and involved 20 universities and research centers in five countries—the United States, United Kingdom, France, Germany, and China. The project was funded at $3 billion ($6.5 billion in 2022). Amazingly, the HGP reached its goal in 13 years and was completed under budget.

Two factors were critical in making the HGP a success. First, the Sanger Method was modified so that instead of attaching a radioactive molecule to the base, a fluorescent dye was attached. A different color of dye could be attached to each type of base, which meant that all four bases could be run at the same time on a single electrophoresis gel plate. Each base would show up with a different color tag when the plate was put under fluorescent light. Second, advances in computing made it possible for the dye pattern to be read and huge bits of data to be analyzed and processed by a computer instead of manually.

In 2003, the HGP announced that the project was essentially complete. Ninety-two percent of the human genome had been sequenced, leaving 400 gaps that were mainly in DNA whose function was to protect the ends of chromosomes. By March 31, 2022, the last gap had been sequenced. The genome produced was not from a single individual. Blood for the project had been collected from volunteers in the area around Buffalo, New York, and the published sequence represented a composite that was highly accurate because most of the human genome does not vary from person to person.

The success of the HGP benefited many fields from pharmaceuticals to anthropology. Personalized medicine became a realistic goal (see entry 60). Anthropologists and evolutionists had a new tool to clarify relationships among groups. Agricultural crops could be designed for specific environments. Fetal cells could be examined for genetic defects. Cancer cells could be compared to healthy cells, leading to better treatments.

Commercial gene sequencing companies developed to meet these possibilities, and improvements were made to make the process faster and less costly. When Apple founder Steve Jobs had his genome commercially sequenced, it cost him about $100,000 and took days. Today the same procedure costs several hundred dollars and takes hours. The accessibility of individual genome sequencing is one of the most consequential results of HGP, but with commercialization and accessibility came ethical questions.

When a person has their genome sequenced, they acquire information not just about themselves but information that may be relevant to their biological relatives. If their sequencing reveals the potential for a serious health problem, should they tell their relatives? Will this lead to a form of genetic discrimination in hiring or insurance coverage? What if the parentage of a relative is not what that person believes it to be? Should they be told? How will the benefits of individual genomic sequencing be balanced against the potential loss of genetic privacy? These are questions that Sanger probably never imagined. They remain to be answered by citizens in the twenty-first century.

FURTHER INFORMATION

"Frederick Sanger." Britannica, September 6, 2022. https://www.britannica.com/biography /Frederick-Sanger

Mayo Clinic. "What Is Genomic Sequencing?" YouTube, February 7, 2018. https://www .youtube.com/watch?v=2JUu1WqidC4

National Human Genome Project Institute. "Human Genome Project." August 24, 2022. https://www.genome.gov/about-genomics/educational-resources/fact-sheets/human -genome-project

TED-Ed. "How To Sequence the Human Genome—Mark Kiel." YouTube, December 9, 2013. https://www.youtube.com/watch?v=MvuYATh7Y74

50. First Magnetic Resonance Imaging of a Human, 1977

Magnetic resonance imaging (MRI) is a noninvasive method of examining soft tissue in the body. The first MRI scan of a human occurred on July 3, 1977, at the State University of New York Downstate Medical Center in Brooklyn, New York. Raymond Damadian (1936–2022), the inventor of the first MRI machine, performed the scan on his graduate student, Larry Minkoff. Damadian claimed he wanted to be the first person scanned but that he was too large to fit into the machine.

Raymond Damadian was a talented, overachieving child who attended New York City public schools but also studied violin at The Julliard School and participated in Junior Davis Cup tennis competition. At age 15, he won a full scholarship from the Ford Foundation that he used to attend the University of Wisconsin where he studied math and physics. He then went on to earn a medical degree from Albert Einstein College of Medicine.

Originally, nuclear magnetic resonance (NMR) was used to examine the structure of chemicals. The word "nuclear" in this case refers to activity in the nucleus of an atom, not to nuclear radiation. Nevertheless, the word was dropped from the name of the medical procedure for fear that it would scare away patients.

During his research, Damadian discovered that when potassium ions inside cells were exposed to a magnetic field, they emitted a signal of a consistent frequency that differed from the frequency of potassium ions dissolved in distilled water exposed to the same magnetic field. From this and other observations, he theorized that healthy cells and cancer cells might also show differences when exposed to a magnetic field. On the basis of this theory, he proposed the body MRI scanner in 1969, and with a grant from the National Cancer Institute, he built the scanner that was first used in 1977.

An MRI scanner has four main components: a primary magnet that creates a magnetic field, gradient coils, radiofrequency coils, and a computer system. The physics underpinning the MRI scanner is complex, but it is based on the fact that the body consists of about 65% water and that every water molecule is made of one oxygen atom and two hydrogen atoms.

The nucleus of a hydrogen atom is a single positively charged proton. When exposed to the strong primary magnetic field, most of the hydrogen protons line up parallel to the magnetic field. These are called low-energy protons. A minority line up in the antiparallel direction and are called high-energy protons.

The gradient coils generate secondary magnetic fields along different axes, ultimately enabling the creation of images. The radiofrequency coils create pulses of radio waves of specific frequencies. When hit with waves of the correct frequency, some low-energy protons absorb enough energy to move into a high-energy state. This changes their alignment within the magnetic field.

When the radiofrequency pulse is turned off, the low-energy protons that were kicked into the high-energy state by the radiofrequency pulse revert to their low-energy state. This process is called relaxation. The rate of relaxation differs among hydrogen atoms in different types of tissue. These relaxation rates can be measured, and the measurements can be fed into a computer. This process is repeated multiple times. The data are sent to a computer that uses a complex set of algorithms to transform it into an image that shows contrasts among different types of soft tissue or soft tissue abnormalities.

Damadian filed patents on his discoveries and founded a company called Fonar in 1978. The company produced the first commercial scanner in 1980, but hospitals were reluctant to invest in an expensive machine from a company without an established reputation. Fonar continued MRI development and vigorously enforced its patents against other companies that began producing MRI scanners. In one court case, Fonar was awarded $128 million from General Electric for patent infringement. Fonar is now a public company listed on the NASDAQ, and as of 2022, it is still producing innovative MRI scanners. Damadian died from cardiac arrest in August 2022.

HISTORICAL CONTEXT AND IMPACT

The concept of NMR was developed by Isidore Isaac Rabi (1898–1988), the son of Orthodox Jewish parents who immigrated to the United States when he was an infant. Rabi began by studying the magnetic properties of crystals. This evolved into studying the magnetic properties of the nuclei of atoms. Rabi developed a method of precisely measuring these properties. For this he was awarded the Nobel Prize in Physics in 1944. His work served as the foundation for Damadian's development of the MRI scanner.

Damadian was not the only person working on developing MRI. Chemist Peter Lauterbur (1929–2007) became involved with using NMR machines while working in a research lab for Dow Corning Corporation and taking postgraduate classes at the University of Pittsburgh. With this background, when he was drafted into the military, he was sent to work at the Army Chemical Center Medical Laboratory in Maryland. Later he was transferred to a unit that had an NMR machine. This combination of experiences led him to actively consider using NMR to look at biological tissues.

While working at the State University of New York at Stony Brook, Lauterbur developed a different and improved way of using NMR (now called MRI) to create images. The university was so unimpressed with his work that they refused to spend money to file a patent based on it, a decision which, over time, has lost them millions of dollars in licensing fees.

Meanwhile at the University of Nottingham, England, physicist Peter Mansfield (1922–2017) was also working on NMR. After several research stints at various universities, he returned to Nottingham and received funding to build an MRI machine. Mansfield volunteered for the first full body scan using this machine. The scan was performed in December 1978. Mansfield is credited with developing a technique that could localize a specific area of tissue and image it and for improving the speed at which quality images could be made. In 1993, he was knighted and became Sir Peter Mansfield.

In 2003, Lauterbur and Mansfield were jointly awarded the Nobel Prize in Physiology or Medicine. Although Damadian had received many other honors, including a National Medal of Technology in 1988 and the prestigious Lemelson-MIT Lifetime Achievement award in 2001, he was not named along with Lauterbur and Mansfield as a Nobel Prize winner. The Nobel rules allow up to three people each year to receive the Prize related to a new discovery. The rules also specify that the Prize is to be given for a significant new innovation, not simply for the improvement of a technique or product.

Speculation about why Damadian was excluded by the Prize committee suggested that his low-contrast images were not clinically reliable in detecting or diagnosing cancer and that the Prize was being given specifically for magnetic resonance *imaging*, so picture quality was an essential consideration. Damadian was irate. He took out full-page advertisements in prominent newspapers such as the *New York Times*, the *Washington Post*, and the *Los Angeles Times* demanding that the Prize committee reconsider its decision.

The academic community was divided. Some experts felt Damadian's fundamental research and development warranted his being included in the Prize, especially since a single Prize can include up to three people and only two were named. They pointed to his breakthrough paper published in *Science* in 1971 as evidence of the significance of his research. Other equally credentialed academics felt Damadian's work had not gone far enough to be considered definitive and that his images were diagnostically inconclusive.

Underlying the controversy was another, uglier question. Had Damadian's work been undervalued or tainted by the fact that he was a fundamentalist Christian creationist? Damadian had quite publicly affirmed that he believed the Book of Genesis in the Bible told the literal truth about the creation of the Earth in six days and that Adam and Eve were the first humans. This outlook was ridiculed by many of his scientific peers. Although the Nobel Prize committee was asked repeatedly to reconsider the 2003 award, they never did, and the records of their decision-making process will remain sealed until 2053.

Today MRI scans are a standard part of many diagnostic workups. Their advantages over X-rays are that they image soft tissue in great detail and do not expose the patient to any harmful radiation. The clarity of MRI images and their range of use has steadily expanded. Original MRI scanners required patients to lie flat and remain completely still in a narrow, noisy tube for a relatively long period. This can cause severe anxiety and induce claustrophobia for many people. Today there are open non-tubular scanners and multi-position scanners that allow a person to be scanned sitting, bending, standing, or lying flat.

Other improvements include the development of functional MRI (fMRI) that can distinguish between highly active and less active parts of the brain when the patient is performing a specific task, such as looking at a picture or focusing on a specific thought. In addition, perfusion MRI uses a contrast medium to show movement of blood through the brain or heart. It can be used to diagnose a stroke or heart attack. MRI scans are also used as guides to improve the accuracy of surgery.

Prices for MRI scanners typically ranged from $1 million to $3 million in 2022. High-end machines can weigh between three and five tons and can generate a magnetic field strong enough to pick up a full-sized car. As of 2019, approximately 12,000 MRI scanners were in use in the United States where more than 40 million scans are performed each year.

FURTHER INFORMATION

Bobby Jones Chiara & Syringomyelia Foundation. "The Story of the MRI—Raymond V. Damadian, MD." YouTube, February 5, 2019. https://www.youtube.com/watch?v=lVOyerAgNUY

Lauterbur, Paul. "Paul Lauterbur—Biographical." Nobel Prize, 2003. https://www.nobelprize.org/prizes/medicine/2003/lauterbur/facts

Lightbox Radiology Education. "Introduction to MRI Physics." YouTube, September 24, 2013. https://www.youtube.com/watch?v=Ok9ILIYzmaY

Mansfield, Peter. "Sir Peter Mansfield—Biographical." Nobel Prize, 2003. https://www.nobelprize.org/prizes/medicine/2003/mansfield/biographicalOrganization

51. First Baby Born through In Vitro Fertilization, 1978

Louise Joy Brown was the first person born after conception that occurred outside the body. Her birth by planned cesarean section occurred on July 25, 1978, at Oldham General Hospital, Lancashire, England. Brown is often referred to as a "test tube baby," but the procedure that led to her birth, in vitro fertilization (IVF), actually occurred in a Petri dish. Today "test tube baby" is considered a pejorative term.

Lesley Brown, Louise's mother, could not conceive naturally because, although she ovulated normally, her fallopian tubes were blocked so fertilization could not occur. To overcome this blockage, an egg was surgically removed from her ovary using laparoscopic surgery. The egg was fertilized in a Petri dish using her husband's sperm. After the fertilized egg began to divide, it was inserted into the mother's uterus where it implanted and developed normally into a baby girl weighing 5 pounds and 12 ounces (2.6 kg) at birth.

To some people, Louise's birth was a miracle of science, but others rejected the out-of-womb conception and called Louise a "Frankenbaby." Although many in the Roman Catholic Church denounced the method of conception, Cardinal Albino Luciani, who became Pope John Paul I, refused to condemn the procedure. Time would show that Louise was a perfectly normal person. She eventually married and, in 2006, had a son who was conceived naturally.

Robert G. Edwards (1925–2013) was a pioneer in reproductive medicine. He was born in Yorkshire, England, to a working-class family. Scholarships allowed him to receive a university education, and eventually he earned a PhD from University of Edinburgh. After stints at several research institutions, he took a permanent position at Cambridge University. Here he became interested in why a mother's immune system does not reject the fetus as foreign tissue since it contains non-maternal genetic material from the father. Eventually, however, Edwards' interests shifted to in vitro egg maturation. He experimented with mouse eggs and was able to make them mature, only to find that Gregory Pincus (1903–1967) had already done similar experiments with rabbits (see entry 44). At this point, Edwards' interest was in the genetics of early development and not in finding a solution to infertility.

Lacking a source of human eggs to extend his early development research, Edwards went to the United States to work with the doctors Georgeanna (1912–2005) and Howard W. Jones (1910–2015), fertility experts at Johns Hopkins University. This experience turned Edwards' thoughts toward IVF as a way to circumvent infertility. At Johns Hopkins, he was able to make a human egg mature in vitro, but on his return to Cambridge, he still faced the problem of effectively fertilizing the egg. This problem was eventually overcome by making changes to the medium in the Petri dish to make it more sperm-friendly. To progress,

Edwards now needed a supply of experimental human eggs. This is when Patrick C. Steptoe (1913–1988) joined the team.

Patrick Steptoe was a gynecologist at Oldham General Hospital. Although laparoscopy—a procedure that allowed surgeons to see inside the body using minimally invasive surgery—was known in the 1960s (see entry 29), very few surgeons used the technique. Steptoe, however, was an exception who had embraced the technique. Using laparoscopy, he could remove eggs from the ovary of infertile volunteers with minimal surgical impact. This provided Edwards the eggs he needed to continue his work. Edwards and Steptoe performed their first IVF in 1966, but it took another decade to perfect the technique and figure out how and when to transfer the developing embryo back into the mother.

When the news of the infertility team's work leaked out, interest among the popular press was enormous. The researchers faced a choice. They could let the press wildly speculate about the creation of in vitro "Franken" life, or they could speak about their research in an effort to educate the public. They chose to communicate with the media. This brought harsh results.

Medical authorities and colleagues disapproved of speaking to the press, and they, along with many in the public, criticized the team for what they considered inappropriate "human experimentation." After the Nazis' appalling World War II human experiments, this was a particularly sensitive subject. The publicity caused grants for the team's IVF work to dry up. Nevertheless, through Edwards' connection to the Drs. Jones, they obtained funding from a wealthy American philanthropist who had been a grateful patient of the Johns Hopkins doctors. This allowed the research to continue until they finally achieved success with the birth of Louise Brown in 1978.

Jean M. Purdy (1945–1985) was the third essential—but often overlooked—team member. Although she trained as a nurse, Purdy was interested in research more than nursing. Before joining the IVF team in 1969, she had worked in tissue rejection research. She was such a critical part of the team that when she took leave to care for a sick relative, the research stopped until she returned. Purdy was a coauthor on 26 papers with Edwards and Steptoe. She was also the first person to see the fertilized egg that became Louise Brown begin to divide. Regrettably, her life was cut short by malignant melanoma, and she died at age 39.

HISTORICAL CONTEXT AND IMPACT

A successful IVF pregnancy involves removing an egg from the ovary at the correct time, fertilizing it in a Petri dish, and, when the fertilized egg begins to divide, transferring it into the uterus, usually between the third and fifth day after fertilization. Walter Heape (1855–1928), a professor at Cambridge University interested in mammalian reproduction, was the first person to successfully transfer embryos from one female mammal to another on April 27, 1890. He used an Angora rabbit that had been bred the day before and transferred her developing embryos into the fallopian tube of a Belgian hare that had been bred three hours earlier. The Belgian hare gave birth to six healthy offspring. Four were Belgian hares and two were Angoras. This showed that embryo transfer was possible without damaging the mother or offspring.

In 1934 Gregory Pincus (1903–1967), who was interested in how hormones affected the reproductive system, stimulated the development of a rabbit egg in vitro (see entry 44). He did this using hormones rather than rabbit sperm. The egg grew into a parthenogenetic rabbit. Parthenogenetic reproduction, in which no male genetic material is involved, is common in plants, invertebrates, and some birds, but not in mammals. When Pincus's work became public in 1936, the resulting publicity drove him out of Harvard University where he had been teaching. He eventually established his own research foundation and played a crucial role in the development of oral contraceptives.

In 1959, Min Chueh Chang (1908–1991), working at Pincus's research foundation, used IVF of an egg by a sperm to grow a normal rabbit. This showed that out-of-womb fertilization followed by embryo transfer could work. Almost two decades and many attempts later, Louise Joy Brown was born using a similar method.

While IVF research continued in England, in many American states the procedure was illegal. Around 1965, the American Medical Association took a position against IVF and insisted that research must stop. They were opposed by the American Fertility Society headed by Dr. Georgeanna Jones, and research continued but not without problems.

In 1973 in New York City, Dr. Landrum Shettles (1909–2003) agreed to help Dr. John Del Zio and his wife Doris Del Zio have a child by IVF. Fertilization occurred, and the embryo was in a Petri dish in an incubator awaiting transfer into the mother when Shettles's supervisor, who disapproved of IVF, heard of the procedure, removed the Petri dish from the incubator, and destroyed the embryo. The Del Zios later sued Columbia-Presbyterian Hospital where this incident took place and were awarded damages for the destruction of the embryo.

The first IVF baby born in the United States arrived on December 28, 1981. The parents, Judy and Robert Carr, lived in Massachusetts where IVF was illegal. They traveled to Virginia—where the Drs. Jones had opened a fertility clinic in 1980—to have the procedure. In 1987, despite previous approval, the Catholic Church issued a doctrinal statement opposing IVF.

Initial IVF procedures that led to Louise Brown's birth were much less sophisticated than today's procedures. Lesley Brown had to stay in the hospital where blood levels of her hormones were monitored daily. Today urine is used to monitor hormone levels, and although IVF requires many clinic visits, it does not require continuous hospitalization. In addition, Lesley Brown, unlike women today, was not given any fertility drugs to encourage ovulation. Moreover, only one egg was taken from Brown, while today multiple eggs are extracted and fertilized and then the healthiest one or two embryos are transferred to the mother.

The remaining embryos can be frozen for future use. Finally, today's needle laparoscopies are precision-guided by ultrasound images, making egg removal more efficient.

As of 2019, more than 10 million children or between 1% and 3% of babies born worldwide were conceived by IVF. Also, as of 2019, 489 fertility clinics were providing IVF services. Until the U.S. Supreme Court overturned the federal right to an abortion in 2022, IVF remained fully legal in the United States. With control of abortion regulation returned to individual states, the future of IVF remains cloudy because individual states have different definitions of when personhood begins. The legal issues surrounding the future of IVF will likely be decided by court cases.

FURTHER INFORMATION

Assefi, Nassim, and Brian A. Levine. "How In Vitro Fertilization (IVF) Works." TED-Ed. YouTube, May 7, 2015. https://www.youtube.com/watch?v=P27waC05Hdk
Brown, Louise. *Louise Brown—40 Years of IVF*. Bristol: Bristol Books CIC, 2018.
Johnson, Martin H. "Nobel Lecture: Robert G. Edwards." Nobel Prize, December 7, 2010. https://www.nobelprize.org/prizes/medicine/2010/edwards/lecture
Kovacs, Gabor, Peter Brinsden, and Alan DeCherney, eds. *In-Vitro Fertilization: The Pioneers' History*. Cambridge, UK: Cambridge University Press, 2018.
Today. "First 'Test Tube Baby' Louise Brown Turns 40: A Look Back." NBC. YouTube, July 25, 2018. https://www.youtube.com/watch?v=xW5-CrIAwjs

52. First Human Disease Eradicated Globally, 1980

On May 8, 1980, the World Health Organization (WHO) declared the Earth free of smallpox. This was the first—and as of 2022, the only—human disease ever intentionally eradicated. Smallpox is a brutal disease. In the twentieth century alone, it killed an estimated 300 million people.

Smallpox is caused by infection with the *variola* virus. There are two strains of this virus: *v. major* is often deadly; *v. minor* is somewhat more survivable. Smallpox is spread by prolonged face-to-face or close contact with an infected person. Symptoms develop roughly two weeks after exposure and include high fever, body ache, vomiting, and a rash. The rash first develops in the mouth. Once it appears the individual is contagious. The rash spreads to the body and forms fluid-filled blisters. When these blisters full of *variola* virus break, the disease can be spread by direct contact or through clothing or bedding that has absorbed the blister fluid. A person remains contagious until all blisters scab over and fall

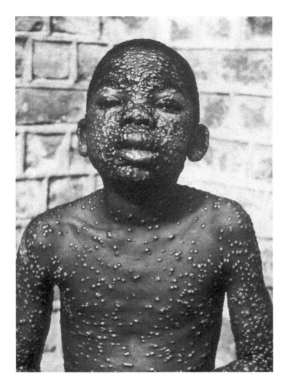

Smallpox was the first disease eradicated through vaccination. The disease was declared eradicated worldwide in 1980. (Centers for Disease Control)

off—a period of about three weeks. About 30% of people who contract smallpox die. Survivors are often left with disfiguring scars and permanent disabilities such as blindness.

Despite the availability of an effective vaccine, in the mid-1960s, 31 countries remained heavily burdened with smallpox, and 10–15 million cases were reported each year. In 1966, the World Health Authority, the decision-making body of the WHO, voted for the second time (the first vote was in 1958, but little funding was provided) to attempt to eradicate smallpox. Many health experts were skeptical that this goal could be reached; however, two factors encouraged them. First, humans are the only species the virus infects. There are no animal reservoirs where the virus can grow and then be passed back to humans. Second, scientists had developed a highly effective vaccine that provided long-lasting immunity if it could be given to a high percentage of the unvaccinated.

At WHO a rumor circulated that the Russians wanted an American to be in charge of the eradication program because they believed that it would fail. Dr. Donald Ainslie Henderson (1928–2016), an American epidemiologist, was selected to head the program. He mobilized a group of 150,000 "pox warriors" from 73 countries to report outbreaks and give vaccinations. The pox warriors faced difficulties in accessing remote locations and overcoming vaccine hesitancy, civil wars, transportation accidents, animal attacks, and illness as they traveled through remote regions. Ethiopian rebels captured one team, but in the end, the pox warriors were successful.

In October 1975, Rahima Banu (1972–), a three-year-old Bangladeshi girl living on Bhola Island in the Bay of Bengal, was the last person in Asia to contract a case of *variola major*. The pox team did not have enough vaccine to vaccinate the entire island, so they used a technique called "ring vaccination" where they vaccinated every living within 1.5 miles (2.4 km) of Banu's family. This created a wall of immunity that kept the disease from spreading. Banu recovered and lived to marry and have children. Ring vaccination is still used. It was successful in containing an outbreak of Ebola virus disease in Africa in 2015.

Ali Maow Maalin (1954–2013), a Somali hospital cook, was the last person to be diagnosed with community-acquired *variola minor*. He was diagnosed on October 26, 1977, after which 50,000 people were vaccinated using the ring vaccination approach, and Maalin's infection did not spread. Maalin recovered, and although he had previously refused vaccination, he became a vaccine worker, persuading Somalis to accept vaccinations against polio and measles. Maalin died at age 58 of complications from malaria.

Community-acquired smallpox had been eradicated by 1978, but there were still a few cases of laboratory-acquired disease. Janet Parker (1938–1978), an employee at Birmingham University Medical School in England, was the last person to die of smallpox. Parker was a medical photographer who worked one floor above a laboratory conducting smallpox research. She became ill on August 11, 1978, and died on September 11. Her mother, who had recently been vaccinated and was caring for her, developed smallpox but recovered. More than 500 people exposed to Parker were quarantined for two weeks, but none became ill.

HISTORICAL CONTEXT AND IMPACT

Smallpox is an ancient disease that spread along the trade routes of the ancient world. Traces of the virus have been found in places as diverse as the mummified body of Egyptian king Ramses V (reigned 1150–1145 BCE) and the bodies of sixth century CE Vikings. Various cultures experimented with ways to prevent the disease. The Chinese took dried pox material from someone who had a relatively mild case, ground it into a powder, and puffed the powder up the nose of a nonimmune person. The hope was that the person would get a mild case of smallpox and then develop lifetime immunity. Similar approaches were tried in India and Central Asia. In what is now Sudan, until the early 1900s, parents would tie a rag around the pox-covered arm of a child with mild smallpox. After the blisters broke, they would transfer the rag to a nonimmune child hoping that child would also have only mild symptoms.

Attempts to protect against smallpox came to Europe by way of the Ottoman Empire. Lady Mary Wortley Montague (1689–1762), wife of the British ambassador to Turkey, observed how women in Turkey used a needle to transfer dried, weakened pox material under the skin of nonimmune children who would often develop only mild disease symptoms. She was so impressed that she had her son treated in the same way. Back in London in 1721, when a smallpox epidemic broke out, Lady Montague convinced Queen Caroline to have some of the royal children inoculated this way. Statistically, this was a good gamble. The death rate from intentional inoculation was about 2% while the death rate from community-acquired smallpox was about 30%.

Cowpox is caused by a virus related to smallpox, but cowpox causes much milder illness. Benjamin Jesty (1736–1816), a British farmer, is now credited with being the first person to intentionally infect individuals with cowpox to prevent smallpox. Jesty did not publicize his experiments. Consequently Edward Jenner (1749–1823), a British doctor, is usually credited as the first person to transfer material from a person with cowpox to a healthy person to immunize against smallpox (see entry11).

Although smallpox was often deadly, resistance to vaccination developed quickly in both England and the United States. When an epidemic of smallpox occurred in Boston in 1721, Cotton Mather (1663–1728), the Puritan minister who instigated the Salem witch trials, preached in favor of vaccination, calling it a gift from God. For his efforts, his house was firebombed. In 1855, Massachusetts became the first state to require proof of smallpox vaccination before a child could attend school. This practice spread to every other state. Routine smallpox vaccination of school children continued in the United States until 1972, when it was deemed unnecessary.

Smallpox has a nasty history of being weaponized. British general Sir Jeffery Amherst attempted to defend Fort Carillon (now Fort Ticonderoga) during the French and Indian War by intentionally providing local tribes supporting the French with blankets infected with smallpox virus. The resulting epidemic decimated entire tribes of native peoples who had no previous exposure to the disease.

Some historians also claim that during the Revolutionary War, the British intentionally introduced smallpox into the Continental army during the 1775 campaign to take Quebec. Whether smallpox was intentionally or accidentally spread, so many Continental soldiers were sickened that the campaign failed. As a result, George Washington, who had survived smallpox as an adolescent, disobeyed explicit instructions from the Continental Congress and forced his soldiers to be vaccinated. By 1777, the death rate of Continental soldiers from smallpox had dropped from 17% to 1%.

The threat of smallpox as a bioterrorism weapon remains today. In 1996, the World Health Authority brokered an agreement that all smallpox virus samples would be destroyed by 1999 except for research samples kept in two high-security laboratories, one in the United States and one in Russia. This was done to reduce the chance of the virus being acquired by terrorists, but it remains unclear how thoroughly the agreement has been carried out. In addition, scientists point out that with genetic sequencing and gene splicing, it is likely that a smallpox virus could be created from scratch in a laboratory. Using mail-order DNA from Germany, a Canadian scientist in 2017 created horsepox, a relative of smallpox that does not harm humans. To protect against a smallpox bioterrorism incident, the United States keeps on had millions of doses of the smallpox vaccine ACAM2000.

Monkeypox, a virus related to smallpox, became a concern in May 2022 when outbreaks of the disease occurred in many Western countries. Monkeypox was first isolated from captive primates in Denmark in 1958. The first human case was identified in 1970 in a child in the Democratic Republic of the Congo. More human infections occurred in Central and West Africa, but in 2003, the disease was found in people outside of Africa. Monkeypox causes blisters resembling those of smallpox. The disease, although thoroughly unpleasant, is rarely lethal. JYNNEOS, a vaccine against monkeypox, is available. Studies have shown, however, that vaccination against smallpox provides about 85% protection against monkeypox, even when the smallpox vaccine was given as long as 60 years ago.

FURTHER INFORMATION

Berman, Joel. "Donald Ainslie Henderson (1928–2016)." *Nature* 24, no. 538 (2016): 42. https://www.nature.com/articles/538042a
"Could Smallpox Ever Come Back?" DECODE by Discovery. YouTube, October 9, 2018. https://www.youtube.com/watch?v=oMhpb8SPe_8
Discovery UK. "The Eradication of Smallpox—Invisible Killers." YouTube, May 6, 2018. https://www.youtube.com/watch?v=dVmkYSkQEN8
Henderson, Donald A. *Smallpox: The Death of a Disease: The Inside Story of Eradicating a Worldwide Killer.* Amherst, NY: Prometheus, 2009.
Tetsekela Anyiam-Osigwe. "The World's Last Smallpox Patient." GAVI, February 10, 2021. https://www.gavi.org/vaccineswork/authors/tetsekela-anyiam-osigwe

53. First Robotic Surgery, 1985

Robotic surgery is surgery carried out using tools attached to a mechanical arm that is remotely controlled by a surgeon with computer assistance. The process is also called robot-assisted surgery. Although the idea of using robots in surgery was conceptualized in 1967, the first robotic surgery on a human did not occur until April 11, 1985. The operation was performed at Memorial Medical Center in Long Beach, California, on a 52-year-old man who wished to remain anonymous.

The first surgical robot was a modified industrial robot known as the Programmable Universal Manipulation Assembly (PUMA) 560. PUMA 560 was designed by American robotics pioneer Victor Scheinman (1942–2016) in collaboration with Unimation, the world's first robotics company. Scheinman's first experience with robots came in elementary school while watching the 1951 movie *The Day the Earth Stood Still*. The movie featured an eight-foot-tall robot that gave Scheinman nightmares. His father, a psychiatrist, encouraged him to build a wooden model of a robot to banish his fears.

Scheinman enrolled at Massachusetts Institute of Technology at age 16 where he studied aeronautics and astronautics. He then spent time working in the defense and space industries. In 1969, as a visiting professor at Stanford University, he designed a special robotic arm, now called the Scheinman Stanford Robot Arm, that was eventually incorporated into the PUMA 560. The robot was originally developed for use by General Motors on its assembly line, in what was one of the early uses of an industrial robot.

Dr. Yik San Kwoh, director of radiology at Memorial Medical Center, developed the software and modifications to make PUMA 560 the first successful surgical robot. Dr. Ronald Young performed the surgery, which involved taking biopsies of three tumors located deep within the brain where they would be difficult to reach by traditional surgical methods. Successful execution of this type of biopsy required years of surgical training, a high level of steadiness, and extreme precision. Even a small misjudgment or tremor in the surgeon's hand could send the needle off course and damage brain tissue.

For this operation, computed tomography scans of the brain (see entry 47) were used to produce a three-dimensional image to pinpoint the tumors. The patient's head was clamped into a special frame that provided location reference points to a computer that calculated several paths through the brain that could be taken by the biopsy needle to reach the tumor. The neurosurgeon chose the optimal path based on his experience, and the robot was programed to execute the surgeon's choice.

The robot's arms were equipped with a variety of tools. First, a small hole was drilled in the patient's skull. Once the opening in the skull was established, the drilling tool on the arm was changed to a biopsy needle. The robot, guided by the surgeon using a trackball controller, then followed the computer-generated path through the brain. When the needle reached the tumor, the surgeon directed the robot to take tiny samples of brain tissue to be analyzed for cancer. The patient was able to leave the hospital within a few days rather than staying a week or longer as would have been required after a traditional brain biopsy.

The first robotic operation was not widely publicized, partly because of the patient's wish to remain anonymous. A second identical operation was performed on a 77-year-old woman at the same hospital by a different surgeon about two months later. This operation received national attention, and many sources referred to it as the first of its kind.

HISTORICAL CONTEXT AND IMPACT

The Czech writer Karel Čapek (1890–1938) created the word "robot" and first used it in his 1921 play *Rossumovi Univerzální Roboti* (Rossum's Universal Robots). "Robot" is a word derived from the Czech word for forced labor, but it soon came to refer to a mechanical device that performed repetitive human tasks. Nevertheless, the definition of "robot" was not formalized until 1979 by the Robot Institute of America.

Initially, some robotics experts were skeptical about the safety of robot-assisted surgery while others were enthusiastic, predicting that robots would revolutionize delicate operations such as repairing blood vessels and damaged spinal discs. The optimists were correct, and after a series of successful operations, robotic surgery was recognized as a technological breakthrough.

Robotic surgery offers many advantages. The robot arm can obtain tissue samples using a smaller hole in the skull than a human surgeon would need. The smaller hole means that the procedure can be done under local anesthesia given in the scalp rather than general anesthesia, since only the scalp, and not brain tissue, feels pain. These advances reduce the patient's hospital stay to one or two days from what used to be a week or more. The steadiness of the robot arm also eliminates any accidental brain damage caused by tremors in the surgeon's hand, and the robot is not affected by fatigue or stress that might be experienced by a human surgeon.

Controlling a robot to biopsy the brain is complicated. Doctors wanting to perform robotic surgery need extensive training even though they are expert neurosurgeons. The surgeons using the PUMA 560 practiced repeatedly by using the computer and robotic arm to locate a single BB pellet embedded in a watermelon. Despite assistance from the robot arm, the surgeon remains in control of the robot and can abort the operation if unexpected events are encountered.

Soon after the success of robot-assisted brain needle biopsies, several companies developed other specialized surgical robots. In 1987, a robot successfully assisted in a cholecystectomy (gallbladder removal). Robots for use in prostate surgery soon followed. One drawback was that each type of operation required a specialized robot, and these robots were expensive. The modified PUMA 560 cost $1 million in 1985, the equivalent of $2.75 million in 2022.

It quickly became apparent that robots were especially well suited for laparoscopic surgery. Laparoscopic surgery uses a laparoscope—a thin, flexible tube equipped with a video camera—that is inserted into the body through a small incision (see entry 29). The camera relays a picture to a monitor where the surgeon can see it. This video technology came into its own in the mid-1980s, and the number of laparoscopic surgeries increased substantially. By 1994, Computer Motion, a company based in Santa Barbara, California, had combined laparoscopic video technology with robotic technology to produce a voice-controlled robotic system for laparoscopic surgery.

In robot-assisted laparoscopic surgery, a laparoscope is inserted into the body through a small incision and relays a picture to a monitor. The robot is armed with the necessary surgical tools, which are inserted through either the same incision or another small incision. The surgeon, guided by the image from the laparoscope, then uses a computer and manipulates the robot arm to perform the surgery.

Several things make robots especially well suited for laparoscopic surgery. First, the robot arm can hold surgical tools too small for a surgeon's hand. Second, the robot can move these tools in more dimensions than the human hand and wrist. Human wrists and fingers, for example, cannot bend back to reach behind themselves while a robot's joints can. This allows robots to reach places and perform surgeries that would be awkward for a larger, less-flexible human hand. Third, robots are not limited to two hands and ten fingers, and fourth, unlike humans, robots are immune to distraction. The advantages of laparoscopic surgery for the patient are smaller incisions, less bleeding, and shorter hospital stays.

Another breakthrough in robot-assisted surgery came in 2000 when the Food and Drug Administration approved the da Vinci Surgical System. Da Vinci was the first robot-assisted surgical device approved for multiple types of laparoscopic surgeries. It can be used in thoracic, cardiac, colorectal, urological, gynecological, head and neck, and general laparoscopic surgeries. Before this, each type of surgery required a special robot. As of 2021, more than 7 million surgeries worldwide had been performed using the da Vinci system. Also, as of 2021, it was the only commercially available device of its kind in the United States.

Another development that has come out of robotic surgery is the capability to perform remote surgery. Much of the groundwork in developing remote robotic surgery was done in the 1970s by the National Aeronautics and Space Administration (NASA) and the Defense Advanced Research Projects Agency (DARPA). This research eventually became the basis for a type of robotic surgery called telesurgery, where a surgeon at some distance from the patient controls a robot that performs a surgical operation. NASA was involved in developing telesurgery because astronauts might need remote access to a surgeon in the event of an illness (e.g., appendicitis) or a serious accident while in space. DARPA saw telesurgery as a way to save the lives of wounded warfighters. With a telesurgery system, the seriously wounded could be operated on without the delay of evacuating them to better-equipped hospital facilities far from the battlefield, and surgeons working from a safe environment could serve multiple widely dispersed units in the field.

The first telesurgery was performed in 2001 in what was called Operation Lindberg. French physician Jacques Marescaux (1948–) and Canadian surgeon Michel Gagner (1960–) operating from New York City used telesurgery to remove the gallbladder of a 68-year-old woman in Strasbourg, France.

Although Operation Lindberg proved that telesurgery was possible, there were drawbacks in terms of delay in communications and providing feedback to the surgeon across long distances. However, with improvements in high-speed communications, telesurgery is being reconsidered. It benefits patients in remote or rural areas where there are few surgical specialists or hospitals that are equipped to perform complex surgeries, and it eliminates the need to transport seriously ill patients long distances to receive the treatment they need.

FURTHER INFORMATION

Blakeslee, Sandra. "A Robot Arm Assists in 3 Brain Operations." *New York Times*, June 25, 1985. https://www.nytimes.com/1985/06/25/science/a-robot-arm-assists-in-3-brain-operations.html

Ganga, Maria L. "L. B. Surgeon Uses Robot in Operation." *Los Angeles Times*, April 21, 1985. https://www.latimes.com/archives/la-xpm-1985-04-21-hl-13493-story.html

George, Evalyn I., Timothy Brand, Anthony LaPorta, Jacques Marescaux, and Richard Satava. "Origins of Robotic Surgery: From Skepticism to Standard of Care." *Journal of the Society of Laparoscopic and Robotic Surgeons* 22, no. 4 (October–December 2018): e2018.00039. https://www.ncbi.nlm.nih.gov/pmc/articles/PMC6261744

Parker, Steve. *20th Century Inventions: Medical Advances*. Austin, TX: Raintree Steck-Vaughn, 1998. (for young readers)

54. First FDA-Approved Drug to Treat HIV/AIDS, 1987

Azidothymidine, commonly known as AZT, was the first drug approved by the U.S. Food and Drug Administration (FDA) to treat infection from human immunodeficiency virus (HIV) that causes acquired immunodeficiency syndrome (AIDS). AZT was synthesized in 1964 by chemist Jerome Horwitz (1919–2012) at Wayne State University School of Medicine in Detroit, Michigan, while working on a grant from the National Cancer Institute. AZT was originally an experimental drug that researchers hoped would kill cancer cells. It failed as a cancer drug, never making it past some unsuccessful animal experiments. Twenty-three years later in 1987, it became the first drug approved to treat AIDS.

Jerome Horwitz, an American organic chemist, became interested in microbiology as a teenager and went on to earn a PhD, partly because he loved science and partly—according to his wife—because he did not want to join his father's wholesale poultry business. Horowitz began his professional career working with rocket fuel at Illinois Institute of Technology but made the rather unusual shift to cancer research and became a professor and researcher at Wayne State University.

Around the time that AZT was synthesized, scientists had a limited understanding of the biochemistry of cancer cells. As a result, most anticancer drugs were developed through trial and error. When AZT failed to cure leukemia in mice, Horwitz wrote a paper about the unsuccessful results and then put the drug aside and pursued other projects. Neither he nor Wayne State University thought it worthwhile to patent the drug, never imagining that it would be resurrected 23 years later as a treatment for a widespread viral disease.

Viruses are not like other cells. They cannot reproduce on their own. To multiply, they need to enter a host cell and use some of the host cell's components to make new virons. Eventually the host cell bursts and the new virons enter the bloodstream, where they infect more cells and repeat the process.

HIV belongs to a class of viruses called retroviruses. Retroviruses store their genetic information as a single strand of RNA rather than the usual double strand of DNA. This changes the way HIV reproduces inside the host cell, but the results are the same—thousands of new virons are eventually released. The host cell for HIV is a type of immune system cell called a CD4 T helper cell. These are cells that help coordinate a full immune system response. Consequently, when these cells are infected and destroyed, the immune response is disrupted and suppressed. This makes HIV difficult to treat. The immune system fails, and other serious infections develop. Left untreated, HIV infection progresses to AIDS and is fatal.

AZT belongs to a group of drugs known as nucleoside reverse transcriptase inhibitors (NRTIs). The genetic material in HIV is RNA, but the host cell's genetic blueprint is made of DNA. Once the virus enters the host cell, it uses host cell components to build a double strand of DNA based on the pattern of its own RNA. It then inserts this new DNA into the host cell's DNA where the inserted material then signals the host cell to start producing new virons. AZT interferes with the virus's process of building the needed double strand of DNA by blocking the addition of components to the end of the new DNA chains.

In the mid-1980s, significant numbers of people were dying of AIDS. Under pressure to find a cure, researchers began testing all kinds of drugs with antiviral properties hoping to find one that would stop the virus. The pharmaceutical company Burroughs Wellcome (now part of GlaxoSmithKline) began investigating AZT in 1984. When the drug appeared potentially effective, they filed a patent on it and funded a human clinical trial of the drug. Under public pressure to do something to stop the AIDS epidemic, the FDA fast-tracked AZT and approved it in March 1987.

Jerome Horwitz and Wayne State University received no money from the patent or the commercialization of AZT, even though they had created the drug. Burroughs Wellcome did donate $100,000 ($250,000 in 2022) to the university, an amount many considered woefully inadequate because a year of AZT cost patients $8,000 ($36,599 in 2022) and Burroughs Wellcome was making a huge profit from the drug.

AZT was not a cure for AIDS, but it did prolong the life of those who were infected. It also triggered a search for other antiviral drugs that could disrupt or suppress HIV reproduction.

HISTORICAL CONTEXT AND IMPACT

Although we tend to think of AIDS as a pandemic disease of the late twentieth century, HIV had been circulating and causing AIDS at a low level in Central Africa for years. The disease had simply not caught the attention of first-world scientists and had not been identified.

AIDS is a zoonotic disease, meaning that it originated in an animal population and later acquired the ability to infect humans. Genetically, HIV is closely related to a virus called simian immunodeficiency virus that infects apes and monkeys. The modern AIDS outbreak is thought to have originated between 1910 and 1920 in what was then Léopoldville in the Belgian Congo (now Kinshasa, Democratic Republic of the Congo). Researchers believe that hunters acquired the infection either by eating the meat of infected chimpanzees or by becoming contaminated with infected blood through breaks in the skin. After that, the disease spread through sexual activity, improper sterilization of needles during vaccination campaigns and drug use, and later through blood transfusions.

AIDS reached the Caribbean around 1971. By 1976, it was found in New York City and San Francisco. Initially, the confusing but deadly symptoms of AIDS were found mainly in men who had sex with men and in intravenous drug users who shared needles. Gradually, the symptoms of this unidentified disease spread to other segments of society. The Centers for Disease Control and Prevention (CDC) marks the official start of the AIDS epidemic as June 5, 1981, although the disease was not named AIDS until August 1982, and it took until 1983 to isolate the virus that caused AIDS symptoms.

In May 1983, French researchers Françoise Barré-Sinoussi (1947–) and Luc Montagnier (1932–2022), working at the Pasteur Institute in France, announced that they had isolated the virus that they believed caused AIDS, although the new virus was not named HIV until May 1986. Barré-Sinoussi and Montagnier shared the 2008 Nobel Prize in Physiology or Medicine for their discovery.

The first commercial test for HIV was approved by the FDA in 1985, but no treatment was available until AZT was approved in 1987. AZT could help suppress AIDS symptoms, but AZT alone could not cure AIDS. HIV mutates rapidly, as do many viruses, and soon the HIV virus developed resistance to AZT. Other drugs in the same class as AZT (NRTI drugs) were tested against the virus, and researchers determined in 1995 that a second drug, dideoxycytidine (also called zalcitabine), taken in combination with AZT slowed the progression of AIDS and reduced the risk of death in people who were asymptomatic but started the drug soon after they tested positive for HIV.

The two-drug combination worked better than AZT alone, but the combination became less effective as HIV continued to mutate and develop resistance against both drugs. In the mid-1990s, new classes of antiretroviral drugs that acted in different ways were developed. This resulted in the development of a triple-drug combination known as highly active antiretroviral therapy or HAART.

As of 2022, there are six classes of antiretroviral drugs amounting to more than 30 approved medications. HAART uses three drugs from at least two different classes that inhibit HIV reproduction in at least two different places in its life cycle. For example, one class of drugs may make it more difficult for HIV to enter into host cells. A second may prevent RNA from acting as a template to make viral DNA, while a third may make it difficult for the new viral DNA to integrate itself into the host cell's DNA. The exact choice of drugs depends on effectiveness, other medications the individual takes, and the side effects the HAART drugs cause.

Ideally HAART is started as soon as the individual tests positive for HIV and before symptoms of AIDS occur. HAART does not cure AIDS, but the three-drug regimen, often combined in a single once-a-day pill, suppresses the virus to a level where it cannot be transmitted to another person. It also helps keep the immune system strong because fewer CD4 T helper cells are destroyed. Many people with HIV infection can lead normal lives as long as they take the HAART drugs. And it all started in 1964 when Jerome Horwitz synthesized AZT looking for a cure for leukemia.

FURTHER INFORMATION

Centers for Disease Control and Prevention. "HIV." September 15, 2022. https://www.cdc
    .gov/hiv/default.html
Human Progress. "Barré-Sinoussi & Montagnier: The Scientists Who Discovered That
    HIV Is the Cause of AIDS—Heroes of Progress Episode 10." YouTube, January 27,
    2021. https://www.youtube.com/watch?v=BuYO9nkkNDU
Kahnacademymedicine. "Treating HIV: Antiretroviral drugs." YouTube, June 26, 2015.
    https://www.youtube.com/watch?v=GR9d9wrOl5E
Park, Alice. "The Story Behind the First AIDS Drug." *Time*, March 19, 2017. https://time
    .com/4705809/first-aids-drug-azt

55. First Laser Vision Corrective Surgery, 1988

The first laser procedure to correct vision in a healthy human eye occurred on March 25, 1988. Ophthalmologist Marguerite McDonald (1951–) performed the surgery at the Tulane University Primate Vivarium in Covington, Louisiana. McDonald was part of a research team from Louisiana State University (LSU) that included Charles R. Munnerlyn (1940–) and Stephen L. Trokel (1938–). The surgery McDonald performed is called a laser photorefractive keratectomy (PRK), and it is still occasionally used today.

McDonald operated on Alberta Cassidy, a 62-year-old woman who had cancer in the bone surrounding her eye. To treat Cassidy's cancer, the eye had to be removed even though it was healthy. Because of this, Cassidy volunteered to let the LSU eye team try the laser PRK technique they had been developing on the eye that was scheduled for removal.

The cornea is the outer, transparent layer of the eye. When light passes from one medium into another (e.g., from air into water), the angle of the light is bent. This bending is called refraction. The cornea is shaped like a dome, and it refracts light passing through it. The light then passes through the pupil (the dark center of the eye) and through the lens beyond it, where a lesser degree of refraction takes place. After this, light hits the retina at the back of the eye. In the retina, special cells convert the energy from light into nerve impulses that the optic nerve carries to the brain where the signals are processed into what we perceive as an image.

Aberrations in the cornea alter the degree of refraction that occurs when light passes through it. This changes where the light hits the retina and affects the quality of the image the individual sees. If the cornea has too much curve, making the dome too pointed, the individual will be nearsighted (myopic). If the cornea is too flattened, the individual will be farsighted (hyperopic). If the cornea is irregular instead of perfectly smooth, the individual will have astigmatism. Early laser PRKs could not correct for astigmatism, only for nearsightedness or farsightedness.

The LSU team had done extensive and often discouraging work perfecting their laser PRK technique before performing Cassidy's procedure. They learned how to use the laser by practicing on thousands of plastic discs. Once they could precisely control the laser, they moved on to practicing on animal cadaver eyes and then on human cadaver eyes. Human cadaver eyes gave the team experience working with eyes of the proper size and shape, but they could not tell the researchers anything about the healing process, the amount of vision change, or any complications from the surgery. To study healing, the researchers used live rabbits. They soon discovered that as the rabbits healed, their eyes developed thick, white scars. This discouraged some of the support staff so much that they decided the project was hopeless and quit. Eventually, however, the scar problem was solved through modifications of the laser to make it function more smoothly.

Laser PRK research had progressed to practicing on the eyes of live monkeys when Alberta Cassidy volunteered to allow the team, with special approval from the Food and Drug Administration (FDA), to use their experimental procedure on her. By the time FDA permission was received, the window had narrowed. Cassidy was scheduled to have the eye removed quite soon, so the laser surgery was performed at the vivarium where the monkey surgeries took place rather than in a hospital.

During the procedure, the epithelium, or top layer of Cassidy's cornea, was removed. Epithelial cells grow back quickly after the operation. Once the epithelium was removed, a specialized laser was used to permanently reshape the underlying stromal tissue. By reshaping the cornea in this way, researchers believed that they could correct a major cause of nearsightedness or farsightedness and give a person normal vision without their needing to wear glasses.

The results of Cassidy's surgery were positive. Healing was uncomplicated, and the visual results were good. Based on this success, the team got permission from the FDA to cut short the monkey experiments and jump ahead to performing laser PRK on the eyes of people who were already blind. With more success, they received permission to use the technique on volunteers with normal vision, and eventually the FDA approved the procedure. Regrettably, Cassidy died from her cancer a short time after her experimental operation, but the eye team persuaded LSU to name a laser laboratory after her in appreciation of her willingness to advance laser eye surgery.

When the LSU team published journal articles on their success and spoke publicly about the procedure, they were met with substantial resistance from the ophthalmology community. Some of the most influential ophthalmologists wrote scathing articles about the procedure, especially after there were reports that vision after the surgery did not always remain stable and up to half of all patients developed an unintended shift toward farsightedness. The opponents of laser PRK objected to the procedure saying it was frivolous, unnecessary, and possibly unethical. Poor eyesight caused by an abnormal corneal refraction was not, they argued, a disease, and the condition could be corrected without the risks associated with surgery simply by wearing glasses. But the procedure was legal, and it gradually gained acceptance.

HISTORICAL CONTEXT AND IMPACT

Refractive keratectomy, or altering the shape of the cornea to change the way it refracts light, was not a new idea. The theory of cutting the edge of the cornea to correct vision dates to an 1896 paper by Lendeer Jans Lans, an ophthalmologist in the Netherlands. In 1930, Japanese ophthalmologist Tsutomu Sato (1902–1960) made delicate cuts in the cornea to change its shape and improve the vision of military pilots. Initially the pilots' vision improved, but soon their corneas began to degenerate, so the surgery was abandoned. The idea, however, was not forgotten. In 1963, Jose Barraquer (1916–1998), a Spanish-born ophthalmologist who established a clinic in Bogota, Columbia, developed a different technique to reshape the cornea for vision correction. He removed a layer of cornea, froze it, reshaped it, and then implanted it back into the cornea. This procedure was both difficult and imprecise and did not catch on. It took the development of the laser to make refractive keratectomy realistic and reliable.

The physics behind the laser (an acronym for "light amplification by stimulated emission of radiation") was proposed by Albert Einstein (1879–1955) in 1917. He mathematically described how when a photon (a unit of light energy) of the correct frequency hits a responsive electron, it will bump the electron up into a higher (excited) energy state. When the electron de-excites, or returns to its normal energy level, it will emit a second photon having the same frequency and characteristics as the initial photon that excited it. This concept provided the theoretical basis for the laser, but it took another 43 years before one could be built.

Theodore Maiman (1927–2007) built the first successful laser in 1960 at the Hughes Research Laboratory in Malibu, California. His design was based on theoretical work by Charles Townes (1915–2015), who won the 1964 Nobel Prize in Physics, and Arthur Schawlow (1921–1999). In its simplified form, a laser consists of an energy source such as electricity or light that can excite electrons in a target material. When the excited electrons de-excite, they give off energy in the form of photons that are identical to the initial exciting energy. If the photons are enclosed in a tube with a mirror to repeatedly reflect them, the process is amplified until a beam of photons all of the same wavelength and all moving in phase in the same direction is created. This uniformity is what makes a laser light beam different from regular visible light, which consists of photons of varying wavelengths that do not move in synchrony.

Other researchers immediately set about improving the initial laser. The first use of a laser in medicine was in dermatology, where it was used to remove tattoos and birthmarks. The excimer laser, which was invented for use in the manufacturing of microchips, was modified for use on human tissue. This laser uses cool ultraviolet light beams that do not burn tissues. The excimer laser was used in the first PRK, and newer iterations of it are still used today.

PRK had certain drawbacks, mainly that it takes several days for the cornea's epithelial cells to grow back and during this time the individual has hazy vision. Laser-assisted in situ keratomileusis or LASIK solved this problem in 1991 by using a laser to cut a flap in the cornea. The flap removes the epidermal cells and a thin layer of the underlying stromal cells. The excimer laser, which is computer-controlled, then vaporizes part of the remaining stroma to correct the shape of the cornea, after which the flap is replaced. As a result, recovery is quick and almost painless. Vision usually improves within 24 hours.

Both PRK and LASIK are performed today. PRK offers some advantages to people whose cornea is too thin for LASIK; however, most people prefer the faster healing of LASIK. Between 1991 and 2021, about 30 million LASIK surgeries were performed worldwide. However, the number of LASIK surgeries has declined from a high of 1.4 million per year in 2000 to 700,000 in 2018, even though the surgery has proved to be safe and effective.

FURTHER INFORMATION

Alila Medical Media. "Laski or PKR? Which Is Right for Me?" YouTube, January 2016. https://www.youtube.com/watch?v=dKANhIU7Sxk

Lawrence Livermore Laboratory National Ignition Facility & Photon Science. "NIF's Guide to How Lasers Work." Undated. https://lasers.llnl.gov/education/how-lasers-work

Nataloni, Rochelle, "30th Anniversary of Laser Vision Correction." Cataract & Refractive Surgery Today, April 2018. https://crstoday.com/articles/2018-apr/30th-anniversary-of -laser-vision-correction/?single=true

Star Talk. "How Lasers Work, with Neil deGrasse Tyson." YouTube, October 2019. https://www.youtube.com/watch?v=t9jtGHXgQvw

"What Is Photorefractive Keratectomy (PRK)?" American Academy of Ophthalmology, 2022. https://www.aao.org/eye-health/treatments/photorefractive-keratectomy-prk

56. First Mammal Cloned from an Adult Cell, 1996

Dolly the sheep was born on July 5, 1996, at the Roslin Institute at the University of Edinburgh. She was the first mammal cloned from a fully differentiated adult cell, but she was not the first cloned sheep. The first cloned sheep had been born at Cambridge University in 1984, and several other cloned sheep had been born before Dolly at the Roslin Institute. These earlier clones were developed from un-differentiated embryonic cells. Embryonic cells have the ability to develop into many types of specialized cells—nerve, muscle, skin, and so forth. The question scientists at the Roslin Institute wanted to answer was whether the DNA of a mature, fully differentiated mammalian cell had the ability to direct the develop-ment of other types of cells the way embryonic cells did.

To clone Dolly, researchers removed a mammary gland cell from a six-year-old Finn Dorset sheep. This type of sheep has a white face. They then removed an egg cell from a Scottish Blackface sheep. They chose these sheep because their different facial coloring would make it easy to tell which sheep was the biological mother even before DNA analysis, which at that time was much slower and more complicated than it is today.

The cell nucleus contains almost all the cell's DNA. To create Dolly, research-ers removed the nucleus of the egg cell from the black-faced sheep and replaced it with the nucleus of the mammary gland cell from the white-faced sheep, a process called somatic cell nuclear transfer (SCNT). This changed the DNA of the egg cell that would develop into a lamb. The process was not easy. Researchers made 277 unsuccessful attempts before Dolly was born.

Once the modified egg began to divide, it was implanted in the uterus of the surrogate Scottish Blackface sheep who then had a normal pregnancy. When Dolly was born with a white face, it was clear that she was the offspring of the sheep that had donated the mammary gland cell and not the surrogate mother. Her birth proved that a mature differentiated mammalian cell had the ability to direct the development of a complete animal whose body contained many differ-ent cell types.

Ian Wilmut (1944–) and Keith Campbell (1954–2012) led the large team that it took to clone Dolly. Wilmut, the principal investigator on the project, was an English embryologist. His main interest was in the potential for therapeutic clon-ing, that is, developing techniques to engineer cells to produce useful pharma-ceutical products. His work opened up the field of regenerative medicine where, today, cells are being used in clinical trials to repair damaged organs or congeni-tal defects. Wilmut was knighted in 2007 in recognition of his contribution to medicine.

Keith Campbell, an English cell researcher and leader of the Dolly laboratory team, was interested in how cells in an embryo differentiate into different types of mature cells. Working with SCNT of mature cells, he had a breakthrough discovery. He found that if he starved a mature adult mammalian cell of nutrients it would go into an inactive state. When the inactive nucleus of the starved cell was put into an enucleated egg cell at just the right time, the adult cell would reactivate and was then capable of directing the formation of an entire new organism. This breakthrough resulted in the creation of Dolly.

When Dolly was born, she appeared completely normal. An analysis of her DNA, however, showed that her telomeres were shorter at birth than usual. Telomeres are protective end caps on chromosomes. Each time a chromosome divides, a tiny bit of its telomere is lost. This means that telomeres gradually shorten as a healthy animal ages. Although Dolly's shortened telomeres could suggest that she was born biologically older than her chronological age, a multitude of health screenings found no physical signs of premature aging.

Dolly spent her life at the Roslin Institute. She had six lambs (a singleton, twins, then triplets) by normal reproductive breeding. She became infected with a virus that causes lung cancer in sheep and had to be euthanized in 2003 at age six. Her body was given to the National Museum of Scotland where it was preserved and is on display.

HISTORICAL CONTEXT AND IMPACT

The word "clone" originated in 1903 with plant physiologist Herbert J. Webber (1865–1946). It is derived from a Greek word that describes a twig broken off a plant and used to propagate a genetically identical new plant. The word started to appear in science fiction stories in the mid-1940s and early 1950s, although it was usually used imaginatively and not in the way scientists understand the word. John Gurdon (1933–), a British developmental biologist, is thought to be the first person to apply the word "clone" scientifically to animals, which he did in 1963.

John Gurdon's work in developmental biology laid the groundwork for the creation of Dolly and later for the production of embryonic stem cells. Gurdon was born in Hampshire, UK, into a distinguished family that traced its ancestry back to the 1100s. He attended Eton, an exclusive boarding school for boys, where, despite his interest in science, he did so poorly in his first biology class that he was not permitted to take any other science classes. Instead, he was permitted to study Greek, Latin, and modern languages. At Eton in Gurdon's time, these classes were considered suitable for boys from good families who were not too bright.

After leaving Eton, Gurdon spent a year with a private tutor studying biology, his real interest. Through a complex set of parental connections and favors, he was admitted to Oxford University but was told he could not study languages. If he could pass some elementary science exams, he would be allowed to study zoology. Gurdon passed and successfully completed the program in zoology.

After graduation, Gurdon applied to do a PhD in entomology but was rejected. Instead, he did his PhD in developmental biology. His advisor suggested that he work with SCNT in a species of clawed frogs. Eventually Gurdon, who was a motivated and persistent researcher, was able to perfect the difficult technique of nuclear transfer and prove that a specialized adult cell transplanted into an egg could produce a complete frog.

In the mid-1950s, scientists were not completely convinced that every mature cell in the body contained a complete set of genes. Gurdon's work proved that differentiated adult cells retained a full set of genes and that in an amphibian, a mature cell could be reprogrammed to produce offspring. This led other researchers to attempt to clone mammals and culminated in the creation of Dolly. John Gurdon, the boy who Eton thought was not bright enough to study science, shared the 2012 Nobel Prize in Physiology or Medicine with Japanese embryonic stem cell researcher Shinya Yamanaka (1962–).

When Dolly was introduced to the world on February 22, 1997, she created a firestorm of worldwide publicity and controversy. The Roslin Institute got more than 3,000 calls from the media seeking information and comment in the week after she was introduced. Some people saw in Dolly the potential for therapeutic cloning that could lead to treatments for degenerative conditions such as Parkinson's disease. Others saw cloning as unethical, unsafe, and cruel to animals.

Wild speculation about producing human clones left many people frightened and confused. Since then, several human cloning hoaxes have occurred. The most notable among them was when Clonaid, a supposed human cloning company founded by the Raelian religious cult and based in the Bahamas, announced that the first cloned human baby had been born on December 26, 2002. Genetic testing proved this, and several later claims of cloned babies, to be false. Two monkeys were cloned in China in early 2018 using the same technique that was used for Dolly, but no humans have ever been cloned.

In the United States, bills making human cloning illegal have been introduced multiple times since 1997, but all have failed to become law. Individual states, however, have passed laws restricting or prohibiting the use of human material for cloning, leaving a bewildering patchwork of regulations. International law is equally confusing. In 2005, the United Nations passed a nonbinding resolution calling for a ban on all human cloning. As of 2018, human cloning was banned in some countries (e.g., Canada, Poland, South Africa), was legal in restricted form, usually for medical research, in some countries (e.g., the United States, United Kingdom, Australia, India), and was almost unregulated in many Asian countries.

The United Kingdom allows farm animals to be cloned but not pet cats and dogs. The United States permits both pet and farm animal cloning. Several thousand cloned cattle, swine, sheep, and goats currently live in the United States. Although the Food and Drug Administration has ruled that meat from these animals is as safe to eat as meat from naturally reproduced animals, the process of cloning is so expensive and prone to failure that it is inefficient to clone animals for food. Animals are usually cloned only for research, to preserve certain valuable genetic traits, or to attempt to save a species from near extinction.

FURTHER INFORMATION

Al Jazeera English. "The Scientific Legacy of Dolly the sheep." February 21, 2017. https://www.youtube.com/watch?v=QO2yySOCMQw

"Cloning FAQs." University of Edinburgh Roslin Centre for Regenerative Medicine, Undated. https://dolly.roslin.ed.ac.uk/facts/cloning-faqs/index.html

"The Life of Dolly." University of Edinburgh Roslin Centre for Regenerative Medicine, Undated. https://dolly.roslin.ed.ac.uk/facts/the-life-of-dolly/index.html

New York Times. "The Story of Dolly the Cloned Sheep—Retro Report." YouTube, October 14, 2013. https://www.youtube.com/watch?v=tELZEPcgKkE

Wilmut, Ian, and Roger Highfield. After Dolly: The Uses and Misuses of Human Cloning. New York: W.W. Norton, 2006.

57. First FDA-Approved Botanical Drug, 2006

In 2004, the U.S. Food and Drug Administration (FDA) created a regulatory pathway for the approval of botanical drugs. On October 31, 2006, it approved sinecatechins (Veregen), the first botanical drug for sale in the United States. The FDA considers a botanical drug as one containing plant material, algae, or fungi. To be approved, sinecatechins went through a series of well-controlled clinical trials that took several years. In the United States, sinecatechins is manufactured by ANI Pharmaceuticals under the trade name Veregen.

Sinecatechins requires a physician's prescription. It is a brown ointment intended for external use only. The drug is approved for most people aged 18 years and older to treat external genital warts and warts around the anus. These warts can occur in both men and women who are sexually active. They are usually caused by two types of human papillomavirus (HPV) but *not* by the types of HPV associated with cervical and anal cancer.

The botanical element in sinecatechins is a water extract from green tea leaves from a variety of the plant *Camellia sinensis*. This plant is an evergreen shrub or small tree native to tropical and subtropical regions of Asia. Its leaves and buds are used to make tea—everything from white tea to green tea to black tea depending on how the plant material is processed.

The genus name for this plant, Camellia, was chosen to honor the Jesuit missionary Georg Kamel (1661–1706), a botanist and pharmacist. Born in Moravia (now the Czech Republic), Kamel was the first European to write a comprehensive description of the healing plants of the Philippines, where he served for 19 years before dying from a diarrheal illness. During his lifetime, he was known as a skilled herbalist. Today he is remembered as one of the most significant pharmacists of the seventeenth century and for his publication of a series of treatises on medicinal plants of the Philippines in *Philosophical Transactions*, the journal of the Royal Society of London.

The only active ingredient in sinecatechins is a 15% extract from *C. sinensis*. The extract is a mixture of catechins, chemicals that have been shown to have antiviral activity against HPV. The drug needs to be applied several times daily for it to be effective. It should not be used by pregnant or breastfeeding women or by anyone who is immunocompromised.

HISTORICAL CONTEXT AND IMPACT

Herbs have been used to treat illnesses for thousands of years. Archeologists have evidence that tea from Camellia plants was used 5,000 years ago in China. Some of the oldest information about herbal treatments for specific diseases comes from ancient Egypt, where the Ebers Papyrus, written about 1550 BCE, is the most extensive collection of medical knowledge from this period. There is no doubt that herbs were used to treat diseases in all early civilizations. Even some of the most common drugs used today, such as aspirin and morphine, have their origins in botanical medicine.

The use of plants and plant extracts has been formalized in traditional Chinese medicine and Indian Ayurvedic medicine. Across the African continent, more than 4,000 different species of plants are known to have been used in traditional healing. Knowledge of these plants was usually passed down orally, so few written records exist to show which plants were used to treat specific illnesses. Indigenous people in North and South America also used herbs for healing as well as in religious ceremonies.

Herbs used as medicines are more accepted in Europe than in North America, where they are labeled as herbal medical products. In 1978, a German government group called the Federal Institute for Drugs and Medical Devices Commission E examined botanical medicines. The E Commission, as it is commonly called, was made up of doctors, pharmacists, toxicologists, and research scientists. Its purpose was to determine if herbs sold in Germany as medicines were safe and effective.

The Commission wrote monographs on 380 drugs outlining their benefits, drawbacks, side effects, and safety. About 200 of these drugs received their stamp of approval. Why, then, did it take until the twenty-first century for the FDA to approve the first botanical drug?

The United States makes a distinction in the way it regulates pharmaceutical drugs and dietary supplements. Herbal health products are considered dietary supplements. In the Dietary Supplement Health and Education Act of 1994, the

Congress defined dietary supplements as products intended to be taken by mouth in addition to, but not as a replacement for, food. This includes vitamins, minerals, amino acids, herbs and other botanicals, enzymes, live microbials more commonly called probiotics, and ingredients commonly found in food such as protein powders.

The FDA does not regulate dietary supplements. They are regulated under the same laws as food products. For example, a dietary supplement manufacturer must comply with Good Manufacturing Practice, a system designed to assure consistent quality.

Unlike pharmaceuticals, dietary supplements do not go through rigorous clinical trials to prove their safety and effectiveness before they are marketed. The FDA examines a dietary supplement only after consumer complaints are received about it. For example, ephedra (ma-huang), a widely used weight loss supplement, was removed from the U.S. market for safety reasons in 2004 after there were complaints about it causing heart palpitations, anxiety, and mood swings.

Dietary supplement labels must include information on how the product affects or maintains human body components or functions. Unlike pharmaceuticals, dietary supplements cannot claim to diagnose, cure, mitigate, treat, or prevent a disease. In other words, a calcium supplement can say "helps build strong bones," but it cannot say "prevents osteoporosis."

Unlike dietary supplements, approved botanical drugs may be manufactured in any form from tablet to powder to injectable liquid and can be sold either over the counter or by prescription. Normally, the first step in the FDA-approval process is a toxicology study to assure the botanical is safe for humans. If the plant source has a history of being safely used in traditional medicine, this step may be skipped.

Next, an investigational new drug application is filed, and a series of clinical trials begin. These establish safety, effectiveness, correct dosage, and side effects that will determine if the benefits of the drug outweigh the risks. The process can take years and is expensive.

Studies have found that most botanical drugs fail in the FDA-approval process for three reasons: they do not show effectiveness; inexperience of the drug sponsor with the FDA process creates roadblocks; insufficient funding to complete the process. If these hurdles are passed, botanical drug manufactures must show that they have a consistent and uniform source of plant material, something that can be difficult given the variability of soil and weather conditions. Only then will the FDA grant approval for a botanical to be sold as a drug.

Given the difficulty, cost, and length of the FDA-approval process, many pharmaceutical companies find that bringing a botanical drug to market in the United States is not cost-effective. The market for botanicals is primarily in China, India, and South Korea, where there is high acceptance of herbal medicines. The process for manufacturing botanicals in these countries is much less costly and time-consuming, and the market is larger and more accessible than in North America. This makes FDA approval less valuable. As a result, since sinecatechins was approved in 2006, only one other botanical drug has been approved.

In late 2012, crofelemer (Fulyzaq), manufactured by Salix Pharmaceuticals, became only the second botanical drug to be approved by the FDA. Crofelemer is used to treat diarrhea in HIV/AIDS patients who are receiving antiretroviral therapy. The drug is derived from the red sap of *Croton lechleri*, the Amazonian dragon blood tree. Sap from this tree has been used for hundreds of years in traditional healing as an antiseptic and to slow bleeding.

Although more about 50 pharmaceutical companies have submitted more than 500 investigational new drug applications for botanical drugs to the FDA, as of 2022, no other botanical drugs had been approved.

FURTHER INFORMATION

American Botanical Drug Association. "Definition of the FDA 'Botanical Drug.'" YouTube, June 20, 2021. https://www.youtube.com/watch?v=9oP0ksh5NFs

Hoffman, Freddie Ann, and Stephen R. Kishter. "Botanical New Drug Applications— The Final Frontier." *Herbalgram* 99 (Fall 2013): 66–69. https://www.herbalgram.org /resources/herbalgram/issues/99

United States Food and Drug Administration. "Questions and Answers on Dietary Supplements." May 6, 2022. https://www.fda.gov/food/information-consumers-using -dietary-supplements/questions-and-answers-dietary-supplements

University of Kansas Medical Center. "The History of Some Medicinal Plants from 40,000 B.C. to Modern Drugs." YouTube, January 14, 2013. https://www.youtube.com /watch?v=OtpKkboh1uk

58. First Country to Legalize Euthanasia, 2011

The Netherlands became the first country to legalize euthanasia when the Dutch parliament approved by a two-thirds majority a controversial bill that went into effect in April 2011. Initially the bill allowed terminally ill children as young as 12 years old to request euthanasia without parental consent. This provision led to intense resistance. In the bill that finally was passed, the age at which a minor could request euthanasia without parental consent was raised to 16.

The Dutch bill (often called the mercy killing bill) requires that patients wanting to end their life must be experiencing unremitting and unbearable suffering for which there is no reasonable solution or expectation of improvement. It does not specify that the patient be terminally ill, have a limited time to live, nor does it require a waiting period. Doctors are required to consult another independent physician and follow strict guidelines and highly regulated procedures in order to remain immune from prosecution.

Although the Netherlands was the first country to formally legalize euthanasia, long before legalization in 2011, the state had stopped prosecuting Dutch physicians for honoring voluntary end-of-life requests. By some estimates, euthanasia or assisted death accounted for about 3,000 deaths or 2.3% of all deaths in the Netherlands annually.

The vocabulary around assisted dying can be confusing and is interpreted in different ways by different physicians and legal systems. The word "euthanasia" is derived from two Greek words that translate into "good" and "death." In its original Greek meaning, euthanasia meant a natural death that was quick and relatively painless.

Over time, euthanasia came to be understood to mean the hastening of death to relieve the pain and suffering of the terminally ill. Voluntary euthanasia is conducted by a doctor with the consent of the patient. Nonvoluntary euthanasia is performed when the patient is unable to give consent but the consent is given by someone else on the patient's behalf, usually by a close relative or a court. Often these patients have made their wishes known through living wills or advance medical directives before becoming incapacitated. Involuntary euthanasia, on the other hand, is murder. It is illegal. In many countries suicide and any form of assisted suicide is also illegal.

Euthanasia can be active or passive. Active euthanasia occurs when a lethal substance is administered either by a doctor or by the patient to end the patient's life. Passive euthanasia is open to a variety of interpretations, ranging from refusing or withdrawing medical care to prescribing a drug intentionally or unintentionally that does not kill but that will hasten death. "Assisted suicide" and "assisted death" are common terms that have a variety of legal and common meanings depending on the laws of a country and an individual's understanding of those laws.

HISTORICAL CONTEXT AND IMPACT

Assisted dying has been going on, usually clandestinely, for centuries. Hippocrates of Kos (460–c. 370 BCE) developed what is now called the Hippocratic Oath, a set of medical ethics that physicians were sworn to uphold. The Hippocratic Oath prohibited a physician from performing euthanasia; nevertheless, surviving documents show that it was common for Greek physicians to ignore the Oath. Hemlock, which causes respiratory arrest, was their drug of choice for hastening death.

Euthanasia has been widely condemned based on the Christian, Jewish, and Islamic doctrine that life is a sacred gift from God. For centuries, ending life by euthanasia or assisted suicide has been punishable by both religious and secular law. The practice became a political concern in the early 1800s when the use of anesthetics (see entry 15) created new, painless ways to potentially hasten death. As a result, in 1828, New York became the first state to explicitly outlaw assisted suicide, which was defined as providing a weapon or poisonous drug to anyone intending to take their own life.

In the late 1800s, public debate arose about the ethics of mercifully ending the life of individuals who were suffering and had no chance of recovery. In 1885, the American Medical Association put out a statement specifically opposing euthanasia, claiming that it would turn the physician into an executioner. Despite this, in 1906, bills that would legalize euthanasia were introduced in the Ohio and Iowa state legislatures. Both bills failed, but the idea of hastening death for those with incurable suffering was slowly gaining momentum. In 1936, a bill legalizing euthanasia was introduced in the British House of Lords, and in 1937 a similar bill was introduced in the U.S. Senate. Neither bill passed, but the fact that they were debated indicated increasing public interest in end-of-life options.

The movement to legalize euthanasia ended abruptly during World War II (1939–1945) when the full horror of the Nazi eugenics program that involuntarily euthanized the disabled, the mentally ill, and other "undesirables" as well as other Nazi medical atrocities became apparent. From the end of World War II through the early 1970s, support for euthanasia in the United States declined, and the issue almost disappeared from public debate.

In the mid-1970s, the patients' rights movement brought about a reconsideration of end-of-life treatment. Several states passed "right to die" laws. These allowed patients to refuse treatment or request discontinuation of life supporting care such as a ventilator, but they did not permit active euthanasia. This was not enough for Dr. Jack Kevorkian (1928–2011) whose actions brought publicity and debate to the end-of-life question.

Kevorkian was a pathologist and euthanasia rights activist who by his own admission participated in over 130 euthanasia deaths. His position was that terminally ill patients had the right to choose death and that their death was not a crime. In 1987, he began advertising himself as a physician "death consultant" and performed his first public euthanasia in Michigan in 1990. At the time, the state of Michigan had no laws concerning assisted suicide, so no charges were filed against the doctor.

Kevorkian continued to assist patients who wanted to control their own deaths. His critics claimed that he violated medical ethics by euthanizing patients he had known only a few days, some of whom—autopsies suggested— were not terminally ill. He was also criticized for not having the mental state of the patients independently evaluated. In 1998, on the nationally televised program 60 Minutes, Kevorkian showed a videotape of himself administering a lethal dose of medication to a young man who had amyotrophic lateral sclerosis (ALS, Lou Gehrig's disease). In 1999, a court convicted him of second-degree murder for this act, and he was sentenced to a prison term of 10–25 years, of which he served 8 years.

In the United States, there is a lack of consensus on how to regulate assisted dying. Supporters of assisted suicide argue that the patient has the ultimate right to make an unforced decision to die, that the process relieves pain and suffering, and that it allows death with dignity. Opponents, including the leaders of almost all major religions, see assisted suicide as interfering with God's will and the natural cycle of life. Most opponents equate euthanasia with murder, although they may find acceptable the withdrawal of medical care that will lead to a natural death. As a result of these conflicting views, the regulation of voluntary end-of-life choices has been left up to each state.

In 1994, Oregon became the first state to pass the Death with Dignity Act, which legalized physician-assisted suicide. As of 2022, physician-assisted suicide remains illegal in 40 states and is legal in 10 states. Of those 10 states, 9 require that the physician follow certain protocols that the patient be an adult with six months or less to live, meet state residency requirements, and make at least two oral requests for life-ending treatment at least 15 days apart. Some states also require a written request and consultation with a second physician. In the 10th state, Montana, the legality of physician-assisted suicide was decided by the Montana Supreme Court, and there are no specific legal protocols.

Internationally, at least 10 countries permit some form of assisted dying under specific circumstances. The Netherlands has the most liberal policy. In 2020, in a highly controversial move, it exempted from prosecution doctors who euthanize, with parental consent, severely ill infants and terminally ill children between the ages of 1 and 12 years. Canada controversially liberalized its euthanasia laws in 2022, making it easier for people to commit doctor-assisted suicide.

Physicians have been thrust into the middle of the assisted suicide debate. Just like their patients, physicians have moral and religious views on the topic and may be caught between strongly held personal beliefs and patients' legal right to assisted dying. Some consider hospice palliative care an alternative that makes their participation in assisted suicide unnecessary. As of 2019, the official policy of the American Medical Association (AMA) was that physician-assisted suicide is inconsistent with the physician's professional role. The organization urges a multidisciplinary approach to palliative care that will relieve pain and suffering and encourage the use of hospice care as the end of life nears. The AMA does not, however, *require* physicians to refrain from participating in physician-assisted suicide where it is legal.

FURTHER INFORMATION

Brazier, Yvette. "What Are Euthanasia and Assisted Suicide?" *Medical News Today*, June 22, 2022. https://www.medicalnewstoday.com/articles/182951

Burton, Ollie. "Med School Interviews: Euthanasia & Assisted Dying." YouTube, March 24, 2019. https://www.youtube.com/watch?v=8PMrvagshhQ&t=10s

Hiatt, Anna. "The History of the Euthanasia Movement." *JSTOR Daily*, January 6, 2016. https://daily.jstor.org/history-euthanasia-movement

"Historical Timeline: History of Euthanasia and Physician-Assisted Suicide." ProCon,
    March 29, 2022. https://euthanasia.procon.org/historical-timeline
Kaczor, Christopher. *Disputes in Bioethics: Abortion, Euthanasia, and Other Controversies.*
    Notre Dame, IN: University of Notre Dame Press, 2020.

59. First 3D-Printed Prosthesis, 2012

The first use of a 3D-printed hand prothesis occurred in late November 2012. It was developed through a collaboration between two people 10,000 miles apart who would never have met without the internet. The hand project started when Ivan Owen, a special effects prop maker in Bellingham, Washington, made a 3D-printed claw-like hand for cosplay and posted a video of it on YouTube.

Richard Van As, a South African carpenter who had lost fingers in a table saw accident, happened to see the video. He had been searching for a prosthetic finger but discovered that because it had to be custom-made, it cost thousands of dollars more than he could afford. After he saw Owen's video, he tried unsuccessfully to make a similar device for himself. When he failed, he emailed Owen for help.

The long-distance collaboration continued through email until, after making several unsuitable prototypes, it became clear that they would make no more progress unless they met. Owen fundraised to finance a trip to South Africa and received enough frequent flyer miles from a donor to make the visit possible. Together, Owen and As created Robofinger—a finger customized for Van As.

A mother in South Africa saw a Facebook post about Van As's prosthetic finger and got in touch with him. Her five-year-old son, Liam Dippenaar, had been born without fingers on his right hand as a result of a condition called amniotic band syndrome (ABS). In ABS, the amniotic sack that surrounds the developing fetus is damaged, and the fetus becomes entangled with it in a way that restricts blood flow to some part of the body. In Liam's case, limited blood flow kept the cells in his right hand from developing into fingers.

Liam's mother asked if a hand could be made for her son. Van As and Owen rose to the challenge. The company MakerBot donated two 3D printers, and using them, the men were able to make a customized prosthetic hand for Liam for about $150. They called their creation Robohand and chose not to patent the process.

Robohand contained wrist hinges, knuckle plates, and other parts that were 3D printed. The parts were connected with cables and stainless steel bolts. With Robohand, Liam could pick things up and throw a ball. Best of all, when he outgrew the hand, he could get a new, appropriately sized prothesis at an afford-able price.

In January 2013, the files to print Robohand parts were made freely available worldwide, and soon 3D printing hobbyists were volunteering to make hands for those who needed them. This led Ivan's wife, Jen Owen, psychologist and entrepreneur Jon Schull, and Jeremy Simon, whose company sells 3D-printing equipment and project kits, to found e-NABLE, a not-for-profit organization that connects people needing hands with volunteers interested in making them. Today there are e-NABLE volunteers on every continent except Antarctica, and customized prosthetic hands can be made with volunteer labor and between $15 and $50 in materials. This has changed the lives of thousands of children whose parents could not afford new, customized prosthetics that are needed as often as every six to nine months as a child grows.

HISTORICAL CONTEXT AND IMPACT

Protheses have been used since ancient times. One of the earliest was a wooden toe found on a 3,000-year-old Egyptian mummy. The earliest known hand prosthesis was made of iron. It was described by Pliny the Elder (c. 23–79 CE) almost three centuries after its owner had died. Pliny claimed that it was made for Roman general Marcus Sergius who lost his right hand in battle during the Second Punic War (218–201 BCE).

Most early prosthetic hands were made of metal, and most were for soldiers. In the early 1500s, Bavarian knight Götz von Berlichingen (1480–1562), also called "Götz of the Iron Hand," had an iron prosthetic hand made after he lost his own in battle. The prothesis was so heavy that it had to be attached to his armor for support, but it allowed him to hold the reins and control his horse in battle.

World War I (1914–1918) and World War II (1939–1945) accelerated the need for prosthetic limbs. By the 1950s, cables had mostly been replaced by rechargeable batteries to move the fingers in hand protheses. These were still awkward, often uncomfortable devices, but the development of transistors and plastics soon made them lighter and more flexible.

Prosthetic hands continued to improve as new materials such as titanium were introduced, but the prostheses were still limited in their ability to do delicate tasks. Because they had to be custom-built with meticulous craftsmanship, pros-thetic hands could cost as much as $10,000. This put them out of the reach of many people.

In 2016, researchers created an entire hand that was made completely from soft, flexible materials, including the electronic circuitry used to operate the hand. This improvement allowed users to perform more delicate and precise tasks. In 2021, a team at the University of Maryland announced that they had produced a 3D-printed hand that was flexible enough, fast enough, and had enough tactile feedback to beat level one of the Nintendo video game Super Mario Bros.

The groundwork for 3D printing that made these advances possible was established by Hideo Kodama of Japan. In 1981, he developed the idea of using ultraviolet light to polymerize thin layers of a photosensitive resin. Although he built a machine that he called a "rapid prototyping device," he lacked funding and failed to file a patent in a timely way. In 1986, a French team patented a device similar to the one that became the first 3D printer, but their inability to explain commercial uses for the machine led to a lack of funding and the project was dropped.

American Charles Hull (1939–) is considered the father of 3D printing. He developed a technique called stereolithography (SLA) as a faster, cheaper way to make prototypes. Hull patented his first SLA machine in 1986. The device used ultraviolet light to bond thin layers of plastic to make 3D objects. Early machines cost $300,000 ($812,000 in 2022). Hull expected the machines to be used in industry. He had no idea that his invention would change the world of medicine. The first medical use of 3D printing occurred in the late 1990s when the technique was used to make custom dental implants. For his invention of SLA printing, Hull was inducted into the National Inventors Hall of Fame in 2014.

Since the development of SLA, other 3D printing techniques have been devised. Carl Deckard (1961–2019) developed a technique called selective laser sintering (SLS) and S. Scott Crump invented fusion deposition modeling. These different processes work on somewhat similar principles as SLA. A liquid, a gel, or very fine particles are extruded by a nozzle. As the nozzle moves rapidly across a surface, it ejects a thin layer of particles. A special type of light makes the particles bond and harden. The placement of the particles by the nozzle is determined by a computer program. Eventually a solid three-dimensional object is created. The process can take anywhere from a few hours to a few days depending on the size of the object and the capabilities of the printer.

3D printers have experienced an ever-expanding range of uses because of the evolution of faster, more powerful, and less costly computers and advanced computer-aided design software. These developments have extended the industrial use of 3D printing and also reduced the cost of the technology to bring it within reach of many hobbyists. This has allowed organizations such as e-NABLE to harness the skills and goodwill of people across the globe to make inexpensive prostheses as well as changing the prosthesis industry.

More medical advances are expected to come from 3D printing. In March 2022, 3DBio Therapeutics, a regenerative medicine company, announced that it had produced an ear for a woman who was born with one normal ear and one very small ear. The company scanned the normal ear, took cartilage cells from it, grew them in a gel, and then used 3D printing methods to create an identical ear out of the gel that contained the woman's cartilage cells.

The new ear was surgically implanted, and the cells were expected to continue to grow and produce cartilage. The procedure was cosmetic; it did not change the woman's hearing. At the time this book was written, the new ear had not been in place long enough to know if the process was a long-term success. If it is, it would open up a whole new area of regenerative medicine (see entry 60).

FURTHER INFORMATION

Digital Trends. "How Does 3D Printing Work?" YouTube, September 22, 2019. https://www.youtube.com/watch?v=dGajFRaS834

e-NABLE. "Enabling the Future." Undated. https://enablingthefuture.org/about

TCT Group. "Chuck Hull—Inventor, Innovator, Icon: The Story of How 3D Printing Came to Be." YouTube, December 23, 2013. https://www.youtube.com/watch?v=yQMJAg45gFE

Walker, Andrew. "3D Printing for Dummies: How Do They Work?" *Independent*, March 24, 2021. https://www.independent.co.uk/tech/what-is-3d-printing-b1821764.html

Zuo, Kevin J., and Jaret L. Olson. "The Evolution of Functional Hand Replacement: From Iron Prostheses to Hand Transplantation." *Plastic Surgery* 22, no. 1 (Spring 2014): 44–51. https://www.ncbi.nlm.nih.gov/pmc/articles/PMC4128433

60. Firsts of the Future

In the past, many medical firsts were the result of one person's curiosity that led them to work alone for long hours in a laboratory or hospital with rudimentary equipment, trying and failing, making adjustments, and then trying again until they achieved their goal. Some of these innovators got rich. Others never earned a cent for their efforts, but they all gave humanity the gift of improved medical care.

Today, making a medical breakthrough still involves curiosity, determination, long hours, and dedication, but gone are the days of researchers alone in their laboratory. The development of medical advances has become more sophisticated. A medical first in the twenty-first century is likely to be developed by a team with diverse backgrounds in medicine, genetics, computer science, data management, patient rights, and finance. Experimental equipment can cost millions of dollars and expensive clinical trials may take years to complete. Modern researchers face another barrier that earlier researchers did not—the need to promote their vision to acquire funding for their work.

The era of the lone researcher in his lab is gone, but this maybe this is a good thing. By bringing multiple areas of expertise together, who knows what medical firsts may develop. The following are some areas of active research where the next medical first may occur.

REGENERATIVE MEDICINE

Regenerative medicine is a developing field that seeks to find ways to repair or replace damaged tissues. Stem cells are already used to treat blood cancers and other blood disorders. Now researchers want to extend the use of stem cells to repair or replace solid tissues and organs. The field of regenerative medicine is moving along two paths. The most familiar approach is the use of stem cell therapy. Stem cells are cells that with the proper stimulation can differentiate into multiple types of specialized cells.

Embryonic stem cells come from unused embryos that have been donated by women undergoing in vitro fertilization. For in vitro fertilization, multiple eggs are fertilized and allowed to divide for three to five days before they are frozen for future implantation. However, not every embryo will be implanted to initiate a pregnancy. These very early embryos consist of about 150 cells. The cells of unused embryos have the potential to differentiate into many types of specialized body cells that could grow into tissues to replace or repair damaged body parts.

In the United States, ethical and legal questions surround the use of these cells. People who believe that human life begins at conception equate the use of cells from nonimplanted embryos with murder of what they call a "preborn child." Others see them simply as a collection of cells with the potential to save lives. As of 2022, the use of embryonic stem cells is regulated by individual states. Federal legislation could in the future restrict the use of embryonic stem cells and delay or halt the development of regenerative medicine that relies on them.

Stem cells can also be harvested from amniotic fluid before a baby is born or from umbilical cord blood when the mother gives birth. There are no legal or ethical issues associated with this method of acquiring stem cells. Some stem cells can also be harvested from adults. These cells are less flexible than embryonic or umbilical cord stem cells. Under chemical stimulation, they may be capable of differentiating into limited types of specialized cells.

Researchers hope that one day they will be able to inject stem cells into a body part—for example, a knee where cartilage has worn away—and these cells will differentiate into appropriate new tissue. A form of this regenerative procedure is offered at some clinics although it has not been approved by the Food and Drug Administration (FDA). The FDA warns that this type of stem cell use can have negative side effects, and because these procedures are not FDA-approved, they are not covered by insurance.

The field of regenerative medicine is also exploring another option—bioprinting. Bioprinting is 3D printing of new tissue using a substrate that contains living cells. Specialized cells are taken from the individual and grown in a gel-like substrate (see entry 59). New tissue of the correct size and shape is then bioprinted using the techniques of 3D printing. A surgeon would remove the damaged tissue or organ and replace it with bioprinted material. As this book goes to print, a bioprinted ear has been implanted in a woman born with a deformed ear. It is too early to tell how successful this procedure will be, but it may be a first for regenerative medicine.

Other researchers are experimenting with programmed microrobots that can be inserted through a small incision into the body part where tissue needs to be repaired. The microrobot can expand to about 2 inches (54 mm). It will then bioprint replacement tissue exactly where it is needed. One application under development would use bioprint gels containing muscle cells to print a patch inside the stomach at the point where an ulcer has eroded the stomach lining. A first or science fantasy? Only time will tell.

DIAGNOSTICS

How would you feel if every time you went to the bathroom the toilet performed a urinalysis and assessed the presence of blood in your stool and the health of your gut biome using a special test-strip of toilet paper? Researchers at Stanford University are looking at such a possibility. Another Stanford project is a smart bra that would continuously monitor for the development of tiny breast tumors. One problem that the researchers have yet to overcome is where to put the battery to collect the data from the bra. These projects are part of a trend toward continuous monitoring and diagnosis.

Continuous monitoring is already available to record and display glucose levels in the blood of people with type 1 diabetes and to indicate whether the level is trending up or down. This allows individuals to adjust their insulin or sugar intake quickly to keep their blood sugar in an appropriate range—a definite health advantage. Smart watches can already monitor activity, sleep patterns, and heart rate, but makers of medical diagnostic tools want to go far beyond that.

Projects are underway to establish continuous monitoring of blood pressure and the electrical activity of the heart which would be used to produce a continuous EKG reading. Other researchers are developing wearable devices to continuously monitor electrolytes through instant sweat analysis. Over time, wearable and implantable devices will be able to produce an abundance of up-to-the-minute information about the state of one's body and send messages to the individual or a medical care team when unexpected deviations occur.

The ability to generate continuous information is one thing. Who will have access to the information and how it will be interpreted is another. Obvious privacy issues surround the collection and use of medical information generated by continuous diagnostic tools, but there are other concerns. Some continuous diagnostics under development send a message to the individual when the readings show a deviation from the normal range, even if the deviation is small or infrequent. These messages will indicate that a health problem may be occurring and may suggest consultation with a physician. But data interpretation and messaging are only as good as the algorithm they are based on, and some of these diagnostic tools have not yet found a way to distinguish between significant and casual changes in the data.

Supporters of continuous diagnosis suggest that it will be an important factor, along with genetic profiling, in establishing the practice of personalized medicine. Personalized medicine is the provision of the best medicine at the correct dose for each person based on the individual's genetics and biological characteristics.

Concerns about continuous diagnostics center on false positives that generate messages suggesting a health problem. These messages could create anxiety and lead to unnecessary physician visits. Mental health professionals are also concerned that some people may become obsessed with their health data to the point where it interferes with their life and sense of well-being. On the other hand, some physicians worry that people will forego regular checkups because of a false sense of security provided by these diagnostic tools.

DIAGNOSIS BY ARTIFICIAL INTELLIGENCE

Artificial intelligence (AI) and machine learning experts predict that machines may soon be able to read and interpret X-rays and CT scans, making the process faster and less expensive by taking radiologists out of the process. The FDA has already approved some AI algorithms for use in radiology, but images tagged as abnormal are still reviewed by a radiologist. One study found that AI programs and physicians were equally good at reading normal mammograms, but radiologists were more accurate in diagnosing suspect images. The current situation is somewhat analogous to self-driving cars. AI does a good job, but it does not eliminate the need for a driver, or in medicine an experienced physician.

In Australia, a group of AI experts have attempted to write a program to determine by analyzing CT scans whether an individual will die within five years. So far, their program is about 69% accurate—approximately the same accuracy as when physicians make the prediction. The Australian group is working on increasing the program's accuracy. A program that could predict death based on a CT scan would be quite a first. But how many people would want to know the results?

FURTHER INFORMATION

American Association for Precision Medicine. Undated. https://www.aapm.health/about
Australian Academy of Science. "Printing the Future: 3D Bioprinters and Their Uses." Undated. https://www.science.org.au/curious/people-medicine/bioprinting
"Exciting News about the Future of Healthcare." *Medical Futurist*, 2022. https://medicalfuturist.com/magazine
International Consortium for Personalized Medicine. Undated. https://www.icpermed.eu
International Society for Stem Cell Research. "A Closer Look at Stem Cells: Nine Things to Know about Stem Cell Treatments." Undated. https://www.closerlookatstemcells.org/stem-cells-medicine/nine-things-to-know-about-stem-cell-treatments

Mayo Clinic. "Stem Cells: What They Are and What They Do." March 19, 2022. https://www.mayoclinic.org/tests-procedures/bone-marrow-transplant/in-depth/stem-cells/art-20048117

National Human Genome Project. "Personalized Medicine." October 7, 2022. https://www.genome.gov/genetics-glossary/Personalized-Medicine

Personalized Medicine at the FDA. "The Scope & Significance of Progress in 2021." Undated. https://www.personalizedmedicinecoalition.org

Regenerative Medicine Foundation News. Undated. https://www.icpermed.eu

Index

About the Author

Tish Davidson, AM, is a medical writer specializing in making technical information accessible to a general readership. She is the author of *Vaccines: History, Science, and Issues, The Vaccine Debate, What You Need to Know About Diabetes*, and *Hormones: Your Questions Answered*. She has also contributed to *Biology: A Text for High School Students*. Davidson holds membership in the American Society of Journalists and Authors.